I0804768

NIGHTHAWKS

Insider's Guide to the Heraldry and Insignia of the Lockheed F-117 Stealth Fighter

Patrick Allen Blazek

Schiffer Military History
Atglen, PA

ACKNOWLEDGMENTS

Many people have helped me to reach the point in my career which afforded me the opportunity to write this book. My parents, Lawrence and Irene Blazek, have always been there to encourage me to pursue my education and to set my goals high. To my teachers who shared their knowledge, I say Thank You. I will always remember the time spent with my friends, the late Earl Blaschke and the late Buck Phillips. These two individuals nurtured my interest in Electrical Engineering and Flying.

I owe a great deal of gratitude to my former manager, Jim Fitzgerald, for having the confidence to hire me as an Avionics Flight Test Engineer. As a newcomer to the F-117A Flight Test Program, I became interested in the history of the program and, more specifically, the numerous patches produced since the aircraft's inception. This book evolved from the ideas of two individuals, Fitz and myself, engaged in casual conversation on the history of a F-117A Flight Test patch.

I wish to thank all those individuals who have tolerated my persistence while researching the history and acquiring these patches for my collection. Thanks to Glen Harbison of Lockheed Advanced Development Company (LADC) Flight Test, and to Marilyn Forbes of LADC Purchasing for locating the invoices for many of the patches which I referenced for quantities and dates.

My appreciation goes out to my friends and neighbors, Kimmo, Kristina, Joe, and Debbie, who allowed me the use of their personal computers on which this text was initially composed.

I would like to thank the following individuals from LADC for taking the time to verify the information contained within this book. These individuals were an integral part of the F-117A Program, and my respect for them gives me the highest level of confidence that the information I have presented here is to the highest degree of accuracy.

Red McDaris, Manager of Flight Test and Prototype Development; Jim Fitzgerald, Flight Test Avionics Manager; Dick Meathrell, Flight Test; Al Miller, F-117A Flight Test Avionics Supervisor; Jerry Conner, Phil Head, Jerry Horn and Jim McCain, Flight Test Avionics; Richard Bronson, Operations; John Stanley, Production Manager/Manufacturing; and Gary Wendt, Engineering Manager.

To my wife, Marian, who cared for our new son, Earl, as my writings were nearing completion. I could not have finished without your support and encouragement.

Finally, I would like to thank the following individuals who were part of the reviewing process which approved this text for public release:

Paul Martin, LADC Vice President F-117 Programs; James W. Ragsdale, LADC Director of Communications; Steve Shobert, LADC F-117A Program Security; Lou Dachs and Jim Ingram, LADC Legal Branch; and Avrom Solomin, LADC Electromagnetics Division Manager.

I dedicate this book to all those individuals, and their families, who made sacrifices in order for this aircraft to be designed, built, flight tested, and fielded.

Book Design by Ian Robertson.

Copyright © 1999 by Patrick Allen Blazek.

Library of Congress Catalog Number: 99-86279

All rights reserved. No part of this work may be reproduced or used in any forms or by any means – graphic, electronic or mechanical, including photocopying or information storage and retrieval systems – without written permission from the copyright holder.

Printed in China.
ISBN: 0-7643-0681-2

We are interested in hearing from authors with book ideas on related topics.

Published by Schiffer Publishing Ltd.
4880 Lower Valley Road
Atglen, PA 19310
Phone: (610) 593-1777
FAX: (610) 593-2002
E-mail: Schifferbk@aol.com
Please visit our web site catalog at www.schifferbooks.com
or write for a free catalog.
This book may be purchased from the publisher.
Please include $3.95 postage.
Try your bookstore first.

Contents

"For it is the lot of some men to be assigned duties about which they may not speak...such work is not for every man. But those who accept the burdens implicit in this silent labor realize a camaraderie and sense of value known to few. These memories cannot be stolen. They will last always. Untarnished, ever better."

Col. Larry McClain, USAF
15 July 1979

Preface

Having a respect for history, I was fortunate to be hired into an organization responsible for flight testing the most unique aircraft of recent times—the **Lockheed F-117A Stealth Fighter**. As the first fully operational, stealthy aircraft capable of precision weapons delivery inducted into the United States Air Force's inventory, the F-117A is widely known for its performance during Operations Desert Shield and Desert Storm. Comprising 2.5 percent of the USAF's aircraft deployed in the theater, the F-117A was responsible for destroying 40 percent of Iraqi targets.

The heraldry of the Lockheed F-117A Stealth Fighter began many years prior to its history-making flights over Baghdad. Many patches have been produced during the life of the F-117A Program, with the most publicly known patches produced as a direct result of Operations Desert Shield and Desert Storm.

Because of the vast number of patches produced, I have limited my writings to those patches unique to the program. As the F-117A matured and relocated to Holloman AFB, New Mexico, to become part of the 49th Fighter Wing comprised of the 7th, 8th, and 9th Fighter Squadrons, existing USAF patches were adopted to represent them. The information on these patches has been previously documented by the United States Air Force and will not be included.

Although many of these emblems have appeared in earlier publications on the Stealth Fighter, never before has a more concise and accurate portrayal of the F-117A Program's patches and their history been published. As a bonus, this book contains history and lineage of patches produced during the Product Excellence Program's existence for the Lockheed buildings and organizations. Many of these patches represent organizations contributing to the success of the F-117A. Those patches, where the servicemark "Skunk" was modified, were not sanctioned by Lockheed. After many years, Lockheed was successful in getting the word out, "Don't mess with our Skunk!," and has had few enforcement problems since.

Any inaccuracies presented here are either conscious deletions due to the still-classified nature of the information, or are due to the lack of information available.

As a final note, I would like to thank my former manager, Jim Fitzgerald, who, during casual conversation over this subject, inspired the idea for these writings.

I hope that you find the history as interesting and colorful as the patches presented here.

Patrick Allen Blazek
October 1994

F-117A Flight Test and Combined Test Force Patches

4 - 5/8" BAJA SCORPIONS
This first generation patch was designed by Dick Meathrell and the late Dick Abrams. "The name Baja Scorpions was conceived by Dick Meathrell, F-117A Flight Test Administrator, very early in the F-117A Flight Test Program, the result of an assignment to Dick by Flight Test Director, Keith Beswick. Keith had suggested an organization patch featuring a graduating senior's cap with a performance chart's jagged trend line to designate the program's code name, Senior Trend."

"After enduring much derision for his suggestion, Keith assigned Dick Meathrell the task."

"The F-117A Flight Test group was located in a complex at the south end of the test site. To the rest of the personnel at the test site, they were known as the South Enders. From this was derived Baja, the Spanish word (loosely translated) for south."

A rattlesnake came to mind when Dick Meathrell was searching for a mascot. However, this was decided against because, unlike the radar-eluding F-117A, the rattlesnake provides ample warning prior to striking. Then, one day, a scorpion was found in one of the hangars and captured. Since scorpions strike without warning, this potentially dangerous creature turned out to be a blessing by providing a symbolic mascot.

Dick Abrams, then F-117A Flight Test Manager, provided the design of the scorpion from photos taken of the captured critter. The first design of this patch had "Strikes Without Warning" across the bottom of the patch, which was removed at the request of security. On 03 SEP 81, 500 of the 4 5/8 inch patch were delivered to Flight Test.

Special thanks to Nelson Hicks for trading me this highly coveted original Baja Scorpion patch, and to Dick Meathrell for contributing to the above passage.

3" BAJA SCORPIONS
The diameter of this second-generation Baja Scorpions patch was reduced to 3 inches for the order of 300 delivered to Dick Abrams on 02 JUL 82. This reduction in size was a result of feedback from Flight Test personnel indicating that the 4-5/8 inch patch was too big and inconvenient for display on a flight jacket. F-117A Flight Test personnel originally displayed the Baja Scorpions' emblem on their jackets only while in the workplace. When this patch was originally designed, security placed a restriction on the use of this emblem by Flight Test members while away from their work assignment. The Baja Scorpions patch could not be displayed at gatherings where uncleared family members and general public were in attendance. One of these gatherings was ConeStock, an annual picnic sponsored by the F-117A Flight Test Avionics group, known as the Coneheads. After the public disclosure of the F-117A, 400 additional patches were ordered by Flight Test on 20 DEC 89.

A standard pin, a 5 year service pin, and a 10 year service pin with the Baja Scorpion design have also been produced.

CONEHEADS

There is some confusion as to the origin of the name "Coneheads." Some people say that the name dates back to radar operators in World War II, while others say that it originated back in the days of the F-104 and referred to the aircraft's Radar Engineers. The name was adopted by the F-117A's Avionics Engineers, one of the most unique groups within the program's Flight Test organization, but that's another story.

The original concept of this patch was developed by Len Macy and Rick Wyllie of F-117A Flight Test Avionics with additional input from other members of the Conehead organization.

The shape of the Conehead patch derives its form from the top view of the F-117A. In the days prior to the F-117A's public disclosure in 1988, the only reason this patch was approved by security was because the name "Conehead" suggested the same shape. The symbols appearing on this patch are symbolic of the various, then-classified, systems integrated into the F-117A's avionics suite. The eyeball represents the Infrared Acquisition and Designation System (IRADS), while the cardinal points are symbolic of the modified AN/ASN-136 Inertial Navigation System (INS). The original design had a GBU-10 to represent the weapons system, but the bomb was replaced with a fire burst so as not to compromise security. The feedback control system diagram symbolizes the fly-by-wire flight control system incorporated into the F-117A. The control diagram has an intentional open in the feedback circuit. This open loop commemorates a phrase which Don Simanski, an early Conehead leader, always used. An electrical term used to describe a break in a feedback circuit, Don would say "Let's go open-loop," which translates to "Let's party!"

Even the Hexadecimal software encoding system used by the F-117's Delco computers was depicted on the Conehead patch in the form of "4F." "4F," when converted to its decimal equivalent, is 79. Coincidentally, it was the number of the building which the Coneheads frequented after hours to socialize. Many of the early developmental problems with the avionics were solved in this relaxing atmosphere, free of the pressures and rigid environment of the workplace. It was pointed out that 4F was also the code used by the military during World War II to mark the files of those least medically desirable to draft. This fact made 4F more enticing to the Coneheads, who like to find solutions to problems in the most unorthodox ways. An unexpected coincidence was discovered a few years after the design of the Conehead patch. While converting numbers from Decimal to Octal, another software encoding system, it was found that 79 in decimal converted to 117 in octal.

The first-generation Conehead patch was bordered with a gold colored thread. Approximately 200 of these were produced and distributed among the original Coneheads and other members of Flight Test. A second version of this patch with the gold border was produced to give to the non-technical personnel who supported the F-117A Flight Test Program. In this version, the 4F was removed from the hand in order to avoid a lengthy description of the complicated Hexadecimal system to those without an engineering background. Approximately 200 of this version were ordered. In an effort to distinguish the original Coneheads from those who joined the group later in the program, the gold border was changed to silver, thus creating a third version of this patch. Two sizes of pins were also produced of the silver bordered design.

Special thanks to Mr. Jerry Conner for donating the gold Conehead patch without the 4F. Special thanks to Mark Daniel for trading me the gold Conehead patch with the 4F.

YOU CAN'T FLY WITHOUT SUPPLY

Pride and peer pressure were the driving forces behind the design of this patch. The supply organization supported the Flight Test and Blue Streak efforts. All of the groups within the F-117A Flight Test effort were designing their own patches. It was then that Dan Hutchinson, Sr., approached Tex Boydstun to design a patch for the "Supply Guys." The supply group played an important part in keeping the aircraft flying, and it was for this reason that Tex and Dan were proud to be part of this group. The phrase on the patch summed up their feelings best. The blue lightning bolt signified an association with the Blue Streak Mod Team and Lockheed. Approximately 50 patches of this design were made between 1983 and 1984.

Special thanks to Dick Burton, Tex Boydstun, and Dan Hutchinson, Sr., for this hard-to-come-by patch.

T-38

The intense pride of having the opportunity to fly in one of these aircraft was the driving factor behind Jim "Fitz" Fitzgerald designing this patch. The T-38s were used for F-117A safety and photo chase purposes, and to maintain pilot proficiency. To the delight of the flight qualified test engineers, they were always given the opportunity to fill a backseat. What a privilege it was to fly with the world's most talented test pilots, most of whom were very generous to their GIBs (Guys In Back) with stick time.

The shape and layout of the patch were borrowed from the SR-71 9th SRW patch. The colors were simple, black and white, representing the T-38 operations in the Black World of security. Only 25 of these patches were made in 1985 and distributed to the current contractor/USAF test pilots, and flight qualified engineers working on the F-117A Flight Test Program.

Special thanks to former Lockheed chase photographer Mike Portzen for trading me this patch for my collection.

SCORPION FTE

Created by Flight Test Engineers Dan Fuller and Major Dave Coombs prior to the disclosure of the F-117A, this patch was designed as a group patch for the FTEs working on the aircraft. The scorpion denotes the FTEs as part of the Baja Scorpion test team. The stars represent nighttime, the period in which most testing occurred in the early years. The T-38 was the aircraft used to monitor/chase the F-117A during critical tests which required close visual monitoring. The Black color of the T-38 symbolizes the black program, which could not be acknowledged at the time this patch was designed. Between 21 MAY 85 and 26 JAN 89, 350 of these patches were produced during three different orders.

3" BAJA SCORPIONS - 1000 - 10 SEP 85

This patch commemorates a keynote achievement during the F-117A Flight Test Program, 1,000 accumulated test sorties on aircraft S/Ns 780 thru 784. Ed Stamper designed this patch after Lt. Colonel Don Cornell (The Great Carnac impersonator) flew the 0.4 hour long mission in S/N 784 on 10 SEP 85. The 3-inch Baja Scorpions patch was used with "1000" and the date of the achievement, "10 SEP 85," embroidered on it in gold-colored thread. 350 patches of this design were delivered to Flight Test on 30 JAN 86.

Group photos were also taken to commemorate the occasion of certain accumulated number of test sorties with 1,000 sorties being one of them. A 3-foot by 3-foot piece of plywood was painted with the design of this patch and was included in the photo. On 11 OCT 88, Lt. Colonel Jim Dunn flew the 2000th test sortie in S/N 784 lasting 0.6 hours. Although a patch was never produced for this milestone, the tradition was continued. A group photo was taken with the scorpion board now reading "2000" and "11 OCT 88."

Special thanks to Jay Black and Ron Mosher for the addition of this patch to my collection.

DYNAMICS DUO

This cartoon patch was created by Rick Wyllie and Dan Fuller to commemorate the flutter tests performed on the F-117A. The name "Dynamics Duo" refers to the two F-117As used during these dynamic response tests and their respective pilots. However, some of the Flight Test members of the time like to say that the name applied to two engineers, Stephanie McNay and Bill Anderson, who were specifically responsible for monitoring these tests and analyzing the data.

The #1 aircraft was symbolic for S/N 781 piloted by Hal Farley (Bandit 117). At the time, all Lockheed test pilots had blue flight helmets. The blue helmet in this patch was representative of Hal Farley, F-117A Chief Test Pilot. The Band-Aid on S/N 781's tail fin is symbolic of an event which necessitated this series of flutter tests. During a routine test flight piloted by Jon Beesley on 25 SEP 85, the left-side tail fin departed S/N 781 as a result of flutter/vibration. However, the aircraft was successfully landed to later be repaired and further serve Flight Test for future testing, a tribute to the proficiency and bravery of its test pilot, who received a "Well Done" Award for his efforts.

The #4 aircraft was symbolic for S/N 784 piloted by Maj. Steven A. Green during this series of tests. The 55 represents aircraft operating limit 55 (AOL 55) which defines the operating envelope of the F-117A, while the polygon on S/N 784's tail fin denotes a modified, composite replacement fin designed to increase this operating envelope without flutter. Like Hal Farley, Steve was identified by the color of his helmet, which was painted red.

The telescope is symbolic for engineer Dave Coombs. His call sign being "Eyeglass," Dave's primary function was to monitor critical safety of flight parameters as they were being telemetered to the control room. The lines coming from the house, or the control room, are indicative of the burst of data provided from the excitation of each F-117A's Flight Control System when simulating flutter.

Special thanks to Vic Bender and Bob Piasecki for trading me this patch.

SCORPION FLIGHT TEST

This patch was one of the original Flight Test patches and was designed by Dan Fuller and Major Dave Coombs. The airplane was generic for the undisclosed F-117A. The black color symbolized the secrecy of the test program, while the orange wing tips were symbolic of the Flight Test Program. The lines protruding aft of the trailing edges represent contrails. On the nose of the aircraft is a boom with Alpha and Beta probes, which are typically used in the early stages of a flight test program. Between 07 FEB 86 and 26 JAN 89, 955 of these patches were produced during four different orders.

FLIGHT TEST INSTRUMENTATION GETS THE JOB DONE

This design was originally used as a symbol for the Lockheed Flight Test Instrumentation group. The poster version of this emblem, designed by Wayne Maynard, had a number 51 on the shield on the Skunk's chest to indicate the building in Burbank where the instrumentation group resided. When Don Kershisnik converted this design to a patch, the number 51 was changed to a lightning bolt to symbolize the F-117A electrical instrumentation group. Approximately 100 were made from 1986 to 1987. The orange background is representative of the orange color used to identify Flight Test Instrumentation wiring and hardware.

The round patch with the triangular wave are symbolic of the traces seen on the round oscilloscope screens during verification of the instrumentation equipment. A variation of this design used on plaques given to retiring instrumentation personnel utilizes a sine wave instead of the triangular wave.

The Skunk is the registered trademark of the Advanced Development Projects Division of Lockheed, otherwise known as the "Skunk Works." This division of Lockheed Martin Corporation is now known as Lockheed Martin Skunk Works.

Special thanks to Ray Cuilty for donating this patch to my collection.

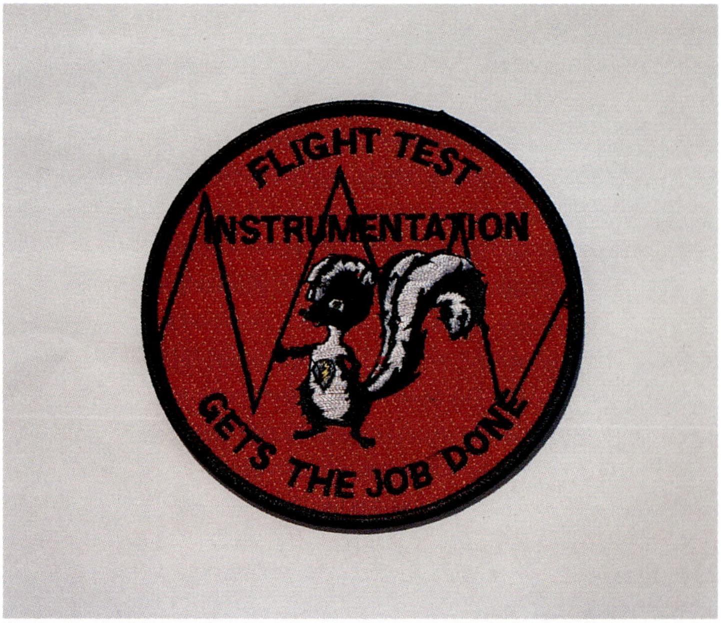

HI AOA

Dan Fuller designed this emblem for the high angle of attack tests flown by test pilots Dave Ferguson (Bandit 105) and Jon Beesley (Bandit 102). The scorpion symbol, a smaller version of the Baja Scorpion patch, subtly reveals a connection with the F-117A. The "HI AOA" lettering was laid out to be symbolic of the then-classified shape of the F-117A. The cloud signifies the secrecy under which these tests were performed.

The "BALS check" is an acronym for the systematic preflight checks made on the following items: **B**attery pack (press to test and set to auto) for the emergency hydro system, **A**nnunciator panel clear of warnings, **L**imiter switch set to Alpha test value desired, and additional **S**traps checked on the ejection seat.

The ball with the roman numeral III and chain were symbolic of the governing body of three who determined the flight safety criteria for these High Alpha tests. This group was known by some as "Troika."

Special thanks to Nelson Hicks and Byron Harmon for donating this patch.

PRODUCTION * FLIGHT

The motivating person behind the design of this patch was Bob "Junior" Clayton, Production Manager of Final Delivery Flight. Junior felt a need to have a patch that was dedicated to and represented those who got their hands dirty working on the production F-117A aircraft. The hand has a tattoo of a Baja Scorpion on it to symbolize connection to the Flight Test group. The wrench and the multimeter test leads are symbolic of the mechanics and electricians, also known as tweaks, who worked on the delivery of each aircraft. The cloud represents the cover of secrecy under which the aircraft operated before its disclosure. The flag, as seen on other patches, is present because of the patriotism felt by those closely associated with the F-117A Program. The design of this patch was drawn by Rick Wyllie. 300 of these patches were made and given to those who were members of the production group.

Special thanks to Tex Johnson and Nelson Hicks for supplying this patch.

SYSTEMS ENGINEERING

This patch was created by Rick Wyllie, at the request of Vic Bender, as a group patch for the Systems Engineers. These engineers were responsible for the following F-117A systems: Armament, Flight Controls, Electrical, Landing Gear, Brakes, Nose Wheel Steering (NWS), Hydraulics, Emergency Power Unit (EPU), Auxiliary Power Unit (APU), Propulsion, and Fuel. The six stars are symbolic for the six engineering categories under which the above systems were divided. The red and white stripes represent the patriotism of the Systems Engineering group. Pins were also produced with this design.

Special thanks to Jerry Horn for supplying this patch for my collection.

MOD BLOCK 1

Approximately 500 patches of this design were ordered by Russ Phillips in 1987. This patch signified the Mod Block 1 modifications performed on the Infrared Acquisition and Designation System (IRADS). Mod Block 1 consisted of a major overhaul to the IRADS' turrets, which included updated boards, heavier stops, routing changes in wiring, etc. The black glasses represent the undisclosed, black F-117A. The two eyeballs on the end of springs have dual significance. The eyeballs themselves are symbolic of the IRADS' two turrets. The bloodshot characteristics symbolize the many hours of hard work performed by nine Texas Instruments and two Lockheed Field Service Reps to incorporate the modifications.

Special thanks to Jerry Conner for donating this patch to my collection.

F-117A 18 JUN 81

This design was created by Dick Abrams to be used as the F-117A CTF/Flight Test patch after the public disclosure of the F-117A Program by the USAF on 10 NOV 88. The original design for this patch incorporated the dates for the 1000 and 2000 test sortie milestones, in addition to the first flight date. However, this version was not approved at the time.

At the time, the Combined Test Force (CTF), comprised of Lockheed Advanced Development Company and USAF personnel, was responsible for the flight testing of the F-117A and its Operational Flight Program (OFP) updates.

The date which appears on this design is significant because it was the date when actual flight testing of the F-117A began. On 18 JUN 81, Lockheed Test Pilot Hal Farley climbed into the cockpit of S/N 780 and performed the first flight ever of a F-117A Stealth Fighter. This flight lasted 0.3 hours and was the beginning of a new era in aviation. Remembering the very first, grainy photograph of the F-117A released by the Pentagon, one would notice that the silhouette of the F-117A on the patch was taken from the photo. Between 14 MAR 89 and 10 SEP 90, 2,400 of these patches and 1,450 pins were produced. This is more than any other F-117A Flight Test patch.

SHIP 4 SCORPION

USAF Master Sergeant Byron Harmon designed this emblem as a patch for the crew who maintained Flight Test aircraft S/N 783. S/N 783, a WDC avionics configured aircraft, was the fourth F-117A built by the Lockheed Skunk Works. Because 783 is also known as Ship 4, Byron began with the number 4 and designed the rest of this patch around it. The shape of this patch was inspired by the Conehead patch and the 9th SRW SR-71 patch. The angles which form the borders of this patch are symbolic of the faceted aircraft. The red and black color scheme, along with the scorpion, were borrowed from the Baja Scorpion patch and denotes that the maintenance crew are members of the F-117A Flight Test Team. The first 50 of these patches were ordered for Byron by a friend of his on duty in Korea in 1989. As with many of the Flight Test patches, word of the Ship 4 patch spread beyond those associated with its intended use. As a result, the original 50 were not enough to satisfy the demand for this emblem. A few months later, 200 more were produced.

F-117A FLIGHT TEST

This patch is a modified version of the original gray and black F-117A CTF/Flight Test patch. Lt. Col Steve Green, CTF Commander, adopted this design as the official squadron patch in 1990 for the USAF members of the CTF. The words "Flight Test" replaced the first flight date of 18 JUN 81. On 03 DEC 90, 200 gray and black patches were delivered. In the same order, 200 subdued patches were also delivered with olive drab and black colors. One may notice that the subdued version is actually brighter than most subdued military patches. The colors did not turn out as intended and, as a result, this subdued version was not very popular amongst CTF members.

LOCKHEED F-117A STEALTH FIGHTER - FLIGHT TEST

This patch was designed by LADC F-117A Flight Test Engineer Glen Harbison to replace the existing gray and black F-117A CTF Flight Test patch. Since the Air Force adopted a version of the original gray and black patch design as their squadron patch, Flight Test members felt that a new patch should be designed to represent the Lockheed Flight Test organization involved in testing the F-117A.

The company's registered trademark, the Lockheed Star, makes an appearance in this design to further symbolize the association of Lockheed and the Flight Test Program. This patch was designed well after the program's public disclosure, and it was not necessary to have any hidden symbols or meanings incorporated into this design. From 07 JUN 91 to 24 SEP 91, 999 of these patches were produced.

CLIMATIC TEST TEAM - F-117A 1991

Aircraft S/N 824 went through preliminary acceptance at Tonopah, Nevada, prior to arriving at Eglin AFB in Florida on the 15th of July 1991 for climatic testing. The purpose of this test was to subject the aircraft to the extreme limits of climatic exposure as defined by the military standard requiring this test and to validate the F-117A's compliance to perform during this exposure. The patch was designed by Ed Harvischak and Cliff Tompkins, who were members of the USAF team assigned to assist with this test that concluded on 20 JAN 92.

The orange border was derived from the color typically used to identify instrumentation wiring and equipment. In essence, the orange color pays tribute to the effort and months of work done in preparation to capture data during this important test. The effects of the sun's heat were reproduced with the use of solar lamps. The polar bear lounging on the iceberg holding a thermometer in one hand and fanning himself with the other was borrowed from Eglin's Climatic Lab patch. This symbol represented the extreme cold which S/N 824 was subjected to during the refrigeration phase of this test. The cloud, lightning bolt, and rain are symbolic of additional elements which S/N 824 was exposed to during another test phase. The snow flurry cloud signifies the snow and ice phase. Two orders for this patch were placed, totaling 350 patches.

Special thanks to Mark King for trading me this patch.

F-117A FLIGHT TEST 3000 30 JUL 92

This patch commemorates a major milestone achieved during the F-117A Flight Test Program, 3,000 accumulated test sorties on S/Ns 780, 781, 782, 783, 784, and 831. USAF Chief Rick Phillips and his crew designed this patch after CTF pilot Maj. Chris Seat flew the 2.1 hour long mission in S/N 784 on 30 JUL 92.

The basic gray and black Flight Test patch was used with "3000" and the date of the achievement, "30 JUL 92," embroidered on it. First, on 11 NOV 92 100 of these patches were received by the CTF, and 50 more were received on 10 MAR 93, due to demand.

F-117 A.G.E. YARD DOGS

This patch was designed by the F-117A Aircraft Ground Equipment (A.G.E.) personnel for the Flight Test organization at Site 7 in Palmdale. In addition to supplying a multitude of ground equipment, these Lockheed employees are responsible for maintaining the equipment. One of the secretaries once said that "The people who work in A.G.E. are like a bunch of junk yard dogs. They're O.K. to have in your yard, but you wouldn't want them in your house." Now, one must remember that a tremendous amount of good-natured rivalry exists between the various groups within Flight Test.

Cartoon characters had been chosen as mascots in the past. The above comment gave the A.G.E. guys an opportunity to choose Grimm from the popular comic strip "Mother Goose & Grimm." Tommy's Heros [*sic*] is in reference to Tom Agnew, who oversees Lockheed's part of these F-117A Flight Test and Depot support personnel. Only 100 of these patches were ever produced. The patches were ordered in October of 1993.

Special thanks to Tex Johnson for donating this patch.

410 TEST SQUADRON

This patch was adopted by the USAF F-117A Flight Test organization located in Palmdale in 1994. The 410th Test Squadron is subordinate to the 412th Test Wing.

"Blue and yellow are the Air Force colors. Blue alludes to the sky, the primary theater of Air Force operations. Yellow refers to the sun and the excellence required of Air Force personnel. The United States flag as a background refers to the unit's support of our national defense. The "X" is a traditional symbol for experimental flight testing and symbolizes the test aircraft the unit flies. The radiating shock waves represent the unknown aspects of test flights and the advance of aeronautical science and technology."[1]

Regular and subdued versions of this patch were produced.

USAF / F-117A / TI / F3 IRADS

This patch was designed by Texas Instruments personnel Roger Smith, Chris Craddock, Jim McElroy, and Dave Gill. The patch was designed to commemorate the achievement of the successful IRADS F^3 (F cubed) program. The program improved the sensor's maintainability, reliability, imaging, and LASER performance. The first 200 of this design were produced in August of 1995, and another 200 were produced in February of 1996.

F-117A FLIGHT TEST 4000 4 NOV 93

Like the previous patch, this patch commemorates a major milestone achieved during the F-117A Flight Test Program. But in this case, the milestone was 4,000 accumulated test hours on S/Ns 780, 781, 782, 783, 784, and 831. This patch was designed after CTF pilot Major Chris Seat flew the 2.0 hour long mission in S/N 831 on 04 NOV 93. On 25 MAR 96, 300 of these patches were ordered by the CTF.

F-117A FLIGHT TEST 5000 12 JAN 96

Like the previous patch, this patch commemorates a major milestone achieved during the F-117A Flight Test Program, 5,000 accumulated test hours on S/Ns 780, 781, 782, 783, 784 and 831. This patch was designed after CTF pilot Major William Gray III flew the 1.4 hour long mission in S/N 831 on 12 JAN 96. 300 of these patches were ordered by the CTF on 25 MAR 96.

F-117A / 18 JUN 81 - 15 YEARS

This patch "commemorates the anniversary of the first flight of one of the greatest advancements in modern aviation warfare technology—THE STEALTH FIGHTER." This flight was flown by Lockheed Test Pilot Hal Farley in S/N 780 on 18 JUNE 1981. On 08 MAY 96, 500 of these patches were received by the CTF.

FLIGHT TEST / 18 JUN 81 - 15 YEARS

This patch commemorates "15 years of Flight Test excellence and innovation by the finest team of visionaries, aviators, technicians, engineers, and mechanics ever assembled, the F-117A Flight Test—Baja Scorpions." On 08 MAY 96, 500 of these patches were received by the CTF.

FLIGHT TEST WEAPONS

The inspiration for this patch came from an emblem designed by USAF MSgt Scott Ashcraft in 1992 and painted by USAF SSgt Yancy Mailes and SMSgt Keith Hilton over the next 3 years. International Insignia produced two sample patches based upon this emblem painted on the wall in the F-117A CTF Weapons' Shop. Upon receiving the sample patches, it was decided to change the color of the scorpion from orange to tan, and to outline the silhouette of the mountains with white. The two sample patches with the orange scorpion remain in the author's private collection.

The CTF had 250 of these patches produced for internal sale in October of 1997. This patch is the first of the Flight Test patches to deviate from the traditional black color of the Baja Scorpion.

MISCELLANEOUS COMMANDS & ORGANIZATIONS
The following Air Force Command and Edwards Air Force Base related patches are all groups that the F-117A CTF have been subordinate to prior to becoming the 410th Test Squadron.

Lockheed F-117A Field Service Patches

BLUE STREAK

Blue Streak is a term typically used throughout the Air Force for fast-action response teams assigned to perform field repairs on aircraft. Blue Streak, in regards to the F-117A, is the Lockheed Depot Field Maintenance Team (DFMT) responsible for performing rapid modifications to the aircraft. These modifications, i.e. configuration package updates and service bulletins, typically took an average of 20 to 30 days to complete.

The Blue Streak team was originally a small organization comprised of approximately 14 people. Blue Streak had collocated with the Flight Test organization and, as a result, the Blue Streakers were the new guys on the block. Not being an official part of the Flight Test group, and not yet having established a reputation for being able to get the job done, the Blue Streak team felt that they were always competing against Flight Test. When trying to co-exist with Flight Test but operate autonomously, Blue Streakers felt "they were always taking it in the rear." This new group on the block needed a mascot, sort of a competition with Flight Test who already had the scorpion. Blue Streak chose Warner Brothers' cartoon character Wile E. Coyote to represent them. Choosing the Road Runner's unlucky nemesis for their mascot sort of epitomized Blue Streak's situation. The blue lightning bolt was merely a symbol for Blue Streak and Lockheed, the company for which the members worked.

These series of patches represent the F-117A Blue Streak team and not the Have Blue Program as depicted in other publications. A Lockheed employee, Jon Chaney, came up with the design of the first-generation patch. In November of 1982, Jon had 100 patches of this design produced. Shortly thereafter, Ron Seres had 40 more

of the 3-inch patches produced. This second order, although very similar in appearance to the first, had slight changes in the coyote's face and ears. The single mountain in these two early versions was representative of Cactus Peak at Tonopah, Nevada, where Blue Streak finally established an operation with the F-117A Fighter Wing.

In August of 1983, Dick Anderson, LADC Management Support Administrator, ordered 100 of a second-generation version that was 4 inches in diameter. In this version, the coyote was improved and more mountains were added. Approximately 400 more of this version were ordered prior to the next change. A 3-inch version of this second-generation patch was produced for ball caps, and it is not known how many of these were produced.

A third-generation, also 4 inches in diameter, was made about 1985. It is estimated that 601 patches were produced from 17 JAN 85 to 08 MAY 86. This third-generation patch had a slightly leaner looking coyote. Unfortunately, the mountains lost some of their definition in this change. The author had 200 copies made of this third-generation patch on 30 JAN 92 because of the demand within the Flight Test organization for a Blue Streak patch.

The DFMT patch was developed during the last days of Blue Streak at Tonopah by Rich Managarin before relocating to Palmdale, California.

The Blue Streak operation was moved to building 602 in Palmdale in 1986. Shortly after the disclosure of the F-117A by the

USAF on 10 NOV 88, the idea of a fourth-generation patch was born. This Blue Streak patch proudly displayed the F-117A in the same view as that of the first-released Air Force photo. Still uneasy about releasing too much information, this design was not approved by security at first. However, with more information being released by the Air Force, Security allowed an unknown number of this patch to be produced for internal use by Blue Streak.

Special thanks to Rich Mangarin for donating the DFMT patch to my collection. Special thanks to Doug Robinson for trading me a first-generation patch. Special thanks to Ron Seres for trading me a second version of a first-generation patch. Special thanks to Jess Arnote for trading me a second-generation patch. Special thanks to Jerry Day for trading me a third-generation patch.

DEPOT TEAM - ACHIEVEMENT DEDICATION PRIDE

The Blue Streak Mod Team moved to Palmdale in 1986, and, as a result, the Depot Team was created. The Depot Center is responsible for all major configuration changes to the F-117A. In 1987, 360 patches were produced, but could not be distributed until well after the disclosure of the F-117A Program by the USAF on 10 NOV 88. Dick Anderson contributed the words "Achievement Dedication Pride" to the design of this patch, drawn by Dennis Quiroz. The first letter from each of these three words are borrowed from the standard abbreviation for the Advanced Development Projects Division of Lockheed where the F-117A was designed and built. The coyote, which had been the on-going mascot for the Blue Streak team, was changed to look more like the famed cartoon character. Wile E. Coyote gives a thumbs up, signifying that everything is O.K., a common signal used by pilots communicating with their ground crew prior to taxiing. The blue lightning bolt is the standard symbol appearing in all Blue Streak related patches.

Special thanks to Tom Agnew and Steve Covate trading me this patch.

F-117A FIELD SERVICE REP / F-117A FIELD SERVICE

This patch was the first of the F-117A Field Service patches designed by Lockheed Field Service Reps. FSRs were assigned to the operational F-117A Fighter Wing in Tonopah, Nevada. The FSRs not only provided real-time technical support to the Wing, but also served as a direct line of communications to the cognizant design engineers in Burbank for problems which could not be immediately solved by the FSRs.

The pride of this responsibility led to the development of this patch originated by Dan Thomas. Doug Ealy, Rod Zarling, Mike Manuel, and Lanny Hentges were some of the FSRs involved in the design and production of this patch. The view of the F-117A depicted on the patch is the same view as the actual F-117A in the photo released during the public disclosure by the USAF on 10 NOV 88. The blue background, moon, and clouds portray the nighttime scenario in which the F-117As operated. The snow-capped mountains are representative of a local landmark in Tonopah, Cactus Peak. The Joshua trees and brown ground characterize the local scenery in Tonopah. The Nighthawk, the symbol and name associated with those assigned to the F-117A Wing, makes an appearance. The red outline of the Nighthawk symbolizes the F-117A's ability to carry out its assigned missions with strength and zeal. The stars are a registered trademark of the Lockheed Corporation. The Skunk is a registered trademark of Lockheed Advanced Development Company and signifies the Skunk Works' involvement in the design, production, and technical support of the F-117A program.

Two versions of this patch were designed in 1989. Between 50 and 100 patches were produced with "Field Service Rep" embroidered on the pennant (scroll) across the bottom of the patch. Only Field Service Reps received this version. Due to the popularity of this patch, a second version was produced for sale outside of Field Service. However, this second version dropped the "Rep" from the pennant across the bottom. This second version is currently available from the original vendor, International Insignia.

Special thanks to Doug Ealy and Marty Roshalt for trading me a first version patch.

F-117A FIELD SERVICE - SAUDI ARABIA

This patch was designed by Doug Ealy and Rod Zarling. Doug and Rod were Lockheed F-117A Field Service Reps assigned to support the 415th Squadron in Saudi Arabia for eight and a half months during Operations Desert Shield and Desert Storm.

The shape of this patch was borrowed from the 415th Squadron patch. The image of a camel and sand was symbolic of the Saudi environment. Lockheed's Field Service personnel in Saudi were represented on the patch by the presence of the Skunk. The view of the F-117A flying out of the patch is a view which Colonel Alton Whitley, Commander of the 37th Tactical Fighter Wing during Operation Desert Storm, approved for use on T-shirts sold in Saudi. Until then, the only view of the aircraft sanctioned for use on patches and T-shirts was the view shown of the F-117A in a banking turn in the first-released photo. The T-shirt, approved by Colonel Whitley, had the new view of the F-117A flying out of smoke coming from a gold magic lantern. The moon on this patch is symbolic of the nighttime operations in which the F-117A participated. However, a small number of this patch was produced without a moon by mistake and later sold.

Approximately 1,500 of these patches were originally produced in March/April of 1991. This design is currently available from the original vendor, International Insignia.

WE OWN THE NIGHT - OPERATION DESERT STORM

Initiated by Mike Manuel and designed by Maria Saetta, this patch represents the Lockheed F-117A Field Service Reps on location in Saudi Arabia during Operation Desert Storm. This patch incorporates the desert and a mosque—scenes symbolic of Baghdad.

The Skunk, the registered trademark for Lockheed Advanced Development Company, represents Lockheed's association with the F-117A. Grim Reapers, a.k.a. the Blue Squadron, was the name of one of the two squadrons of F-117As under the old 4450th. Symbolic of the havoc rendered by the F-117As during Operation Desert Storm, the Skunk takes on the form of the Grim Reaper. A fine example of this havoc was evident during the first few days of Operation Desert Storm. Comprising 2.5 percent of the USAF's aircraft deployed in the theater, the F-117A was responsible for taking out 40 percent of the Iraqi high value strategic targets.

The Skunk drawing a line in the sand is symbolic of President Bush's declaration of a no-fly zone.

Although only 100 were originally produced for issue to the FSRs and members of the 37th Tactical Fighter Wing, a total of 300 were made in 1991. In addition to the patches, 500 T-shirts with this emblem were made. This patch is currently being reproduced by the original vendor, International Insignia.

LOCKHEED LADC FIELD SERVICE

The F-117A Field Service patch was designed by Dave Wells of LADC's F-117A Logistics Field Support Center and Mike Whalen of LADC's Special Arts department. The 4-inch by 4-inch patch was designed to reflect the pride of LADC, the F-117A, and its Field Service operation.

The original F-117A Field Service patch was used by Lockheed Reps assigned at Tonopah, Nevada, to the 4450th Tactical Group and, later, the 37th Fighter Weapons Wing. The relocation of the F-117A Fighter Wing to Holloman AFB, New Mexico, prompted the redesign of the FSR patch in February of 1994.

A total of 108 Field Service patches were delivered by International Insignia of Scottsdale, Arizona, on 15 APR 94.

Special thanks to Dave Wells for trading me this patch and for furnishing the above information.

F-117A NIGHTHAWK

This patch was designed by Dave Wells of LADC's F-117A Logistics Field Support Center in 1994 and ordered by the company to be utilized internally as a F-117A Program patch. Three different sizes of this patch are known to have been produced.

Follow-On Testing and Evaluation Patches

GOAT SUCKERS

If one were to look up the word "Nighthawk" in a dictionary, he would find reference to a synonym for the same bird. This synonym, and the organization's affiliation, are what gave the first operational F-117A training unit its name—Goat Suckers. In anticipation of the USAF receiving its first Top Secret Stealth Fighter, a Replacement Training Unit (RTU) was formed.

A-7s scheduled to be decommissioned were obtained from England AFB in Louisiana and given a home in this newly-formed unit. Pilots who were selected to fly the F-117A flew the A-7s while awaiting the F-117A's arrival in order to maintain their proficiency in a jet aircraft.

In addition to being an appropriate trainer, the A-7s provided an excellent cover story for the pilots when asked what type of aircraft they flew. The cover story became an automatic response from all individuals involved so as to protect the secrecy of the F-117A Program.

Once the training unit obtained its first F-117A, the A-7s were flown by the instructors during chase flights while observing their students who were flying the Stealth.

In January of 1989, the Goat Suckers began replacing the A-7s with Northrop's T-38 Talon.

Special thanks to Mark King for trading me this patch.

PETE'S DRAGON

Some confusion exists as to who was responsible for associating this name to the second Tactical Air Command operational F-117A, S/N 787. However, former Dragon members recall that the name, in part, was borrowed from the Walt Disney character in the movie Pete's Dragon, an invisible dragon which could only be seen by Pete. The name was fitting because the F-117A was invisible, too, but, in its case, to radar.

Reinforcing the choice of the name was the fact that Pete Barnes, TAC's second F-117A operational pilot, was scheduled to receive his first flight in S/N 787 at the time this patch was designed. To Pete's surprise, a green dragon, as seen on this patch, embellished the side of S/N 787 prior to Pete's first flight. A painter for LADC, Brad Brown, had volunteered to perform the job the night before on his own time.

Pete Barnes first flew S/N 787 on 09 JUL 82, and the accomplishment of his first flight in a FOT&E (Follow-On Testing & Evaluation) F-117A was equated to the slaying of a dragon. The USAF Crew Chief during this time was Clyde Fonner. The idea to turn this design into a patch occurred when team members were thinking of ways to raise money for Clyde's retirement party.

Special thanks to Jess Arnote and Ron Mosher for donating this patch.

PETE'S DRAGON II

FOT&E (Follow-On Testing and Evaluation) was responsible for testing the weapons delivery computations of all Operational Flight Programs released from Flight Test. Prior to the Wing receiving an OFP, FOT&E would perform additional weapon deliveries. Up until this time, FOT&E was testing Weapons Delivery Center OFPs with S/N 787, a F-117A configured with the first generation of avionics.

The need for FOT&E to update its aircraft arose with the advent of the second-generation avionics suite known as WSC (Weapons System Computer). S/N 787, a.k.a. Pete's Dragon, was no longer representative of what was being delivered to the Wing from Lockheed's production line.

Lockheed updated the avionics configuration in the middle of production. For each WSC-configured F-117A the Wing received into its inventory, it would send back a WDC-configured F-117A to be updated to WSC by Blue Streak. This means of rotating the F-117As allowed the Wing to maintain an operational number of aircraft, and assured that as many of the F-117As as possible were configured to the current system. Because the Wing was anxious to complete the changeout of its inventory, much effort was required by FOT&E to finally obtain a WSC-configured F-117A.

Acceptance testing was completed on S/N 807 on 28 NOV 84. Shortly thereafter, in the early part of 1985, S/N 807 was assigned to FOT&E and S/N 787 was returned to Blue Streak for update to WSC. To commemorate the occasion of receiving S/N 807, a new patch was designed. Because this was the second FOT&E F-117A, the original patch was modified to the design shown.

Special thanks to Bob Piasecki for donating this patch.

DRAGON TEST TEAM (1ST GENERATION)

By this time the name Dragon had stuck, and USAF personnel associated with Follow-On Testing and Evaluation (FOT&E) began calling themselves the Dragon Test Team. FOT&E decided that they needed a group patch rather than a patch associated with a particular F-117A. The team wanted a patch that was a little more colorful than the earlier Pete's Dragon and Dragon II. The dragon also had to be mean and lean. This required an all new design.

FOT&E was always dropping BDU-33 practice bombs during their testing and evaluation of newly-released Operational Flight Programs (OFPs) from Flight Test. The BDUs were also to be incorporated into the new patch. What you see here is the design which became the first-generation Dragon Test Team logo.

It is believed that this patch was first produced in 1986, and close reproductions are currently available through commercial patch vendors.

DRAGON TEST TEAM - DET 1, 57 FWW
During the Dragon Test Team's first few years of existence, it operated under the jurisdiction of DET 2 of the USAF. The Dragon Test Team members were officially reassigned to DET 1, 57th Fighter Weapons Wing (FWW) out of Nellis AFB in Las Vegas, Nevada, on 05 OCT 89.

Having operated in total secrecy since their formation, the USAF was now publicly recognizing the existence of the Dragon Test Team. The disclosure provided a bit more latitude as to the extent to which the dragons could express their pride. This led to a redesign of their original team patch.

In this new version, the patch became even more colorful, and the dragon became meaner and leaner than in the previous version.

Regular and subdued versions of this patch were produced.

DRAGON TEST TEAM - DET 1, 57 FW
On 01 OCT 91, DET 1, 57th FWW was deactivated and the Dragon Test Team was reassigned as DET 1, 57th Fighter Wing (FW). Later, the Dragon Test Team relocated to Holloman AFB, New Mexico. On 01 APR 93, DET 1, 57th FW was deactivated.

Regular and subdued versions of this patch were produced.

DRAGON TEST TEAM - DET 1, 57 WG
On 01 APR 93, the Dragon Test Team was redesignated as DET 1, 57th Wing (W). DET 1, 57th WG was deactivated at Holloman AFB on 01 OCT 96.

Regular and subdued versions of this patch were produced.

DRAGON TEST TEAM - DET 1, 79 TEG
On 01 OCT 96, the Dragon Test Team was reassigned to DET 1, 79th TEG (Test Evaluation Group). DET 1, 79th TEG is part of the 53rd Wing and remains at Holloman AFB. The checkerboard racing stripe seen in the earlier Dragon Test Team patches was removed in this version.

Regular and subdued versions of this patch were produced.

57th FIGHTER WEAPONS WING
57th FIGHTER WING

On 5 OCT 89, the Dragon Test Team was officially reassigned from the 4450th to the 57th Fighter Weapons Wing (FWW). The 57th FWW was the host command at Nellis AFB, and members of the Dragon Test Team wore this patch along with their unit patch.

The patch was updated when the Air Force dropped the "Weapons" designation in the 01 OCT 91 time frame. Regular and subdued versions of these patches were produced.

Although the 57th FWW was in existence before the F-117A, these patches are included here for reference.

4450th Tactical Fighter Group Patches

TACTICAL AIR COMMAND
At its inception, the F-117A operated under the Tactical Air Command (TAC) of the United States Air Force. The 4450th Tactical Group and the 37th Fighter (Weapons) Wing were TAC organizations which were host to the F-117A Stealth Fighter.

A larger version of this patch was worn by pilots of TAC aircraft prior to 1976. The current, smaller TAC patch and the subdued version were worn by those USAF personnel working on the F-117A Program and are included here for reference.

Later, in June of 1992, TAC merged with Strategic Air Command (SAC) to become Air Combat Command (ACC).

Special thanks to F-117A Pilot Michael R. Cook for giving me this patch removed from his flight suit.

4450th TACTICAL GROUP (RED BORDER)

The design for the red border was the idea of General W. L. Creech, who wanted the red to make the 4450th Tactical Group patch stand out on a flight suit. Because of this patch's red border, it has also been referred to as the official blood patch. The government issue of this patch consisted of a silk-screened emblem with an embroidered edge. A large, 4-inch version was typically worn on the flight suits of the Nighthawk pilots, while a smaller, 3-inch version was issued for use on the left front shirt pocket of fatigues.

Still operating under the cover of the A-7s, the 4450th participated in the 1984 Team Spirit exercise. As is customary when attending an exercise overseas, the 4450th brought with them a quantity of this group patch to trade with their hosts and to use as gifts. This patch was a 100 percent embroidered version of the 3-inch silk-screened blood patch.

Another version of this patch, an unofficial one, soon appeared. This version was a 3-1/2 inch embroidered patch that consisted of the basic emblem without the red border. This patch became a popular item displayed on baseball caps worn by Field Service Reps and other civilian personnel at Tonopah.

"The 4450th's emblem is a Nighthawk in the attack position on a black background, symbolizing night operations. The bird is white, with the bird's eye and lightning bolts embroidered in red to add color."[2]

4450th TACTICAL GROUP COMMANDER'S PATCH

This patch is a compilation of three patches, the 4450th Tactical Group and the "A" and "B" Squadron patches. This patch was originally sold by patch vendors at the Nellis AFB Exchange. Although the designer of this patch is unknown, it is known that it is not an official F-117A patch designed by the 4450th Tactical Group.

Ironically, senior officers associated with the 4450th adopted this patch and wore it on their flight suits. This compilation was worn instead of a single squadron patch in an attempt to promote oneness and to show impartiality within the Tactical Group by senior officers.

NIGHTSTALKERS

The Nightstalkers was a test squadron under the 4450th known as the "A" Squadron. Although the designer of this patch is unknown, it was developed in 1983 for the 4450th "A" Squadron led by Lt. Colonel Ervin C. Sharpe, Jr. On 05 OCT 89, the USAF officially reassigned the unit to the 415th Tactical Fighter Squadron.

The Nightstalkers kept their squadron logo during this transition and added the banner (scroll) with the text "415 TAC FTR SQDN."

GRIM REAPERS

The Grim Reapers was a test squadron under the 4450th known as the "B" Squadron. The official designation of the Grim Reapers was the 4451st Test Squadron. The designer of this patch is also unknown, but it is believed to have been developed in 1983 at the same time as the Nightstalker patch. On 05 OCT 89, the USAF officially reassigned the unit to the 416th Tactical Fighter Squadron.

The patch and motto of the Grim Reapers did not survive the transition to the 416th. It is believed that the Grim Reapers patch did not pass the USAF's patch test of good taste. The 416th is known as the Ghost Riders.

4450th MAINTENANCE SQUADRON (BIG D / A-7D)
This patch was worn on the upper left chest fatigue pocket by the 4450th Maintenance Squadron personnel at Nellis AFB, Nevada. This organization performed the maintenance on the surrogate A-7s flown by the 4450th's pilots. The A-7s were flown by these pilots for flight proficiency during the early days of production, and as a cover story while the F-117A Program remained black. The original A-7 program Big D / A-7D patch was worn by these pilots on their flight suits, while a subdued version was specially produced for use within the 4450th.

Special thanks to Yancy Mailes for trading me this patch.

4461st SUPPORT GROUP

"The blue of the background symbolizes the sky, the Air Force's theater of operation, and the green is the ground beneath it. The target and sword represent the Group's association with TAC. The sword placed over the target symbolizes command and honor. The pyramid, symbol of stability and strength, formed by smaller pyramids, alludes to the individual units of the Group."[3]

This group was later redesignated as the 4450th Combat Support Group (E-Unit) on 24 JAN 89. The Combat Support Group was inactivated on 05 OCT 89, and personnel and assets were transferred to the newly activated Combat Support Group of the 37th Fighter Wing.

Special thanks to Chief Rick Phillips for trading me this patch.

F3XP M.A.R.S.

The induction of the first F-117A, S/N 786, into the USAF inventory brought about some necessary changes. Besides the obvious security issues of operating in total secrecy, the USAF was faced with training its personnel to maintain the stealthy characteristics of this aircraft.

The 4450th quickly established an organization responsible for this maintenance and recruited USAF corrosion control specialists and sheet metal technicians into this classified program, identifying it only by the designator "F3XP." Members of this group, the Materials Application Repair Section (M.A.R.S.), were trained in the techniques of applying and maintaining the coatings used on the F-117A by Lockheed instructors.

Members of this once-secret organization called themselves Martians. This patch was designed by Curt Wallis, one of the early Martians, with their nickname providing the theme.

Pat Burke was the original (Lockheed) factory instructor. The ten day class on these techniques featured physical, hands-on instruction using 4,000 pounds worth of scrap materials unsuitable for flight from Building 199 in Burbank. Approximately one-quarter of the old, wood mock-up was skinned with aluminum and used as a trainer in this class. Students could practice applying coatings to a wing, a rudder, and the aft end of the fuselage forward of the Platy.

The F3XP personnel became so adept with the F-117A that the coatings removed for maintenance were temporarily replaced with painted cardboard to give the aircraft an operational look during unexpected visits from dignitaries. Before the F3XP M.A.R.S. program was formally established in 1984 to 1985, it was known as TFC, The F#&%ing Coating.

Special thanks to Pat Burke for donating this patch.

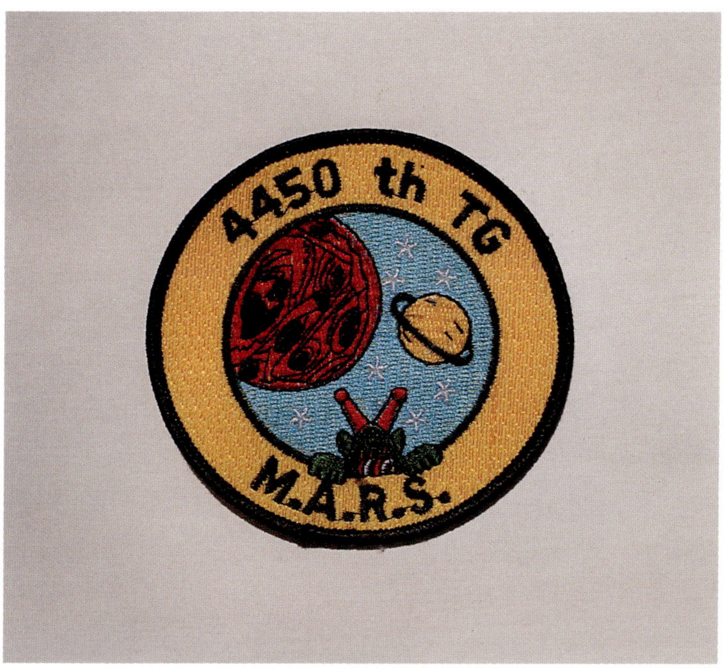

4450th T.G. M.A.R.S.

The previous F3XP M.A.R.S. patch was more or less used by Lockheed and USAF Martians. In 1986, the "F3XP" was dropped and "4450th T.G." was added. This patch became the symbol used by the USAF personnel assigned to the Martian unit. The first order placed was for 150 patches.

Special thanks to Dan Thomas for trading me this patch.

37th Tactical Fighter Wing Patches

37th FIGHTER WING
On 05 OCT 89 the 37th Tactical Fighter Wing moved, in name only, to Tonopah, Nevada, to assume the F-117A mission from the inactivated 4450th Tactical Group. This one act gave birth to a whole new generation of patches.

"A black Nighthawk in flight, poised in an attack position, refers to the Wing's ability with the F-117A Stealth Fighter to meet worldwide commitments with swiftness, strength, and sagacity. The white highlights around the hawk suggest the aircraft's spectral nature, while the white-outlined feathers allude to the F-117A's mastery over a silent approach to its intended target. The background element—a cross of Air Force yellow—denotes the 37th's heritage as the 'Defender of the Crossroads.' The ultramarine blue border between the oriental blue areas and the arms of the yellow cross depicts the 37th's membership in the Air Force and Tactical Air Command community. Moreover, the transition from oriental blue to ultramarine blue insinuates the Air Force's new capability to operate effectively in both day and evening skies. The three red lightning bolts surging from the hawk's claws imply that three operational squadrons are ready to carry out the 37th's mission with strength and zeal for the Air Force. The hawk's red eye denotes its ability to find and track its prey."[4]

Regular and subdued versions of this patch were produced.

37th TACTICAL FIGHTER WING COMMANDER'S PATCH
The history behind this commander's patch is virtually identical to that of the 4450th Tactical Group Commander's patch. This patch is a compilation of three squadron patches and the Wing patch. This patch, too, was sold by the patch vendors at the Nellis AFB Exchange, and the designer of this patch is unknown. However, it is known that the patch is not an official F-117A patch designed by the 37th Fighter Wing.

Nevertheless, senior officers within the 37th adopted this patch and wore it on their flight suits. This compilation was worn instead of a single squadron patch in an attempt to promote oneness and to show impartiality within the Wing by the senior officers.

NIGHTSTALKERS - 415 TACTICAL FIGHTER SQUADRON

"A white Nighthawk, with a red and white halo surrounding its head, symbolizes the F-117A Stealth Fighter. The white, gray, and black feathers suggest the perfection of the F-117A's discretionary weapons delivery capability, its spectral nature, and the maturity of the Stealth mission, while the halo refers to its purity of purpose. The Nighthawk is in an attack position, centered over black mountainous terrain that is separated by a yellow and red sunset from an advancing ultramarine blue evening sky. This signifies the Air Force's ability to operate with strength and zeal in a night environment. A white moon with stratus clouds in the sky denotes the unit's nighttime mission. Red streaks emanating from the Nighthawk's claws refer to its ability to stalk prey."[5]

Regular and subdued versions of this patch were produced. On 01 JUN 92, Strategic Air Command, Tactical Air Command, and Military Airlift Command were disbanded. The 37th became under the control of the newly formed Air Combat Command. As a result, the Air Force did away with "Tactical" squadrons, and, hence, the design was changed. The Aircraft Maintenance Unit assigned to the 415th also produced a subdued patch for its members that had 415 ACFT MAINT UNIT across the bottom of the patch.

GHOST RIDERS - 416 TACTICAL FIGHTER SQUADRON

"A black rider on a black war horse symbolizes the F-117A Stealth Fighter. It also preserves the heritage of the 416th patch. Its white highlights suggest the perfection of the F-117A's weapons delivery capability and the maturity of the Stealth Fighter mission. The horse and rider are in an attack position, centered over a stylized aircraft against an ultramarine blue evening sky. This signifies the Air Force's determined ability to operate with strength and zeal in a night environment. The red of the rider's cape symbolizes the boldness of attacking out of night skies, and the rider's red eyes signify the ability to seek out the enemy in the dark of night. The stylized aircraft represented the squadron's position at the 'Tip of the Spear' of readiness. The white moon and white stars reflect the squadron's nighttime mission. The yellow lightning bolts held in the rider's upraised left arm refers to the squadron's historical excellence and the contribution of the F-117A to tactical deterrence."[6]

On 01 JUN 92, Strategic Air Command, Tactical Air Command, and Military Airlift Command were disbanded. The 37th became under the control of the newly formed Air Combat Command. As a result, the Air Force did away with "Tactical" squadrons and, hence, the design was changed. The Aircraft Maintenance Unit assigned to the 416th also produced a regular and subdued patch for its members that had 416 AMU across the bottom of the patch.

BANDITS - 417 TACTICAL FIGHTER TRAINING SQUADRON

"The 'Ghost' is identical to the 'Ghost' in the historical squadron emblem, with the exception of the mask (for "Bandits"), and suggests an apparition or hard to see being and represents the element of surprise. The black stylized aircraft is representative of the unit's Stealth Fighter mission. The elongated weapons suggest the pin-point accuracy and perfection of the F-117A's discretionary weapons. The descending aircraft (attack position) signifies unit readiness and determination. The moon, cloud, and night environment depict the squadron's primary mission to teach pilots how to employ the ideally suited Stealth Fighter at night. The mountains represent the total environment the 417 TFTS operates in."[7]

On 01 JUN 92, Strategic Air Command, Tactical Air Command, and Military Airlift Command were disbanded. The 37th became under the control of the newly formed Air Combat Command. As a result, the Air Force did away with "Tactical" and "Training" squadrons and, hence, the design was changed. The Aircraft Maintenance Unit assigned to the 417th also produced a subdued patch for its members that had 417 ACFT MAINT UNIT across the bottom of the patch.

37th AIRCRAFT GENERATION SQUADRON

"The emblem symbolizes the 37th Aircraft Generation Squadron's (AGS) direct maintenance support of the F-117A Stealth Fighter. A black Nighthawk, outlined in white, is the sign of the spectral F-117A in action, while four yellow lightning bolts imply that four aircraft maintenance units are providing that support. The white star and moon, accented by an ultramarine blue field, denotes the ability to operate in the sky, the Air Force's primary theater of operations, at night. A black and white target below the Nighthawk represents its ability to find the target in all conditions. A brown, rugged, mountainous terrain falling to a flat plain demonstrates the ability to operate anywhere, while a black-line triangle in the background relates to targeting in the worst of environments. The oriental blue field within the triangle refers to the unit's membership in the Tactical Air Command."[8]

Regular and subdued versions of this patch were produced.

Special thanks to Chief Rick Phillips for donating this patch.

37th CIVIL ENGINEERING SQUADRON

"Three colors are used: ultramarine blue, Air Force yellow, and black. Simple, yet very complex in meaning. Yellow represents the caliber of people working in the squadron—the 'Best of the Best.' Ultramarine blue represents the sky. Our mission, directly and indirectly, is to support our aircraft. Black stands for dignity and the determination to succeed in a mission most are not aware of and the hours of darkness, night, our primary time of operations."

"The symbol of engineering is the castle. Ordinary kingsmen were 'knighted' for outstanding accomplishment. The knight was a symbol of honesty, loyalty, and reliability. They were called upon to accomplish the most difficult tasks. 'The Knights' of Civil Engineering have accomplished the most difficult of tasks while being assigned to a classified location working on classified projects. Thus, the engineering squadron was identified as a 'black organization,' which allowed the term 'Black Knights' to evolve."[9]

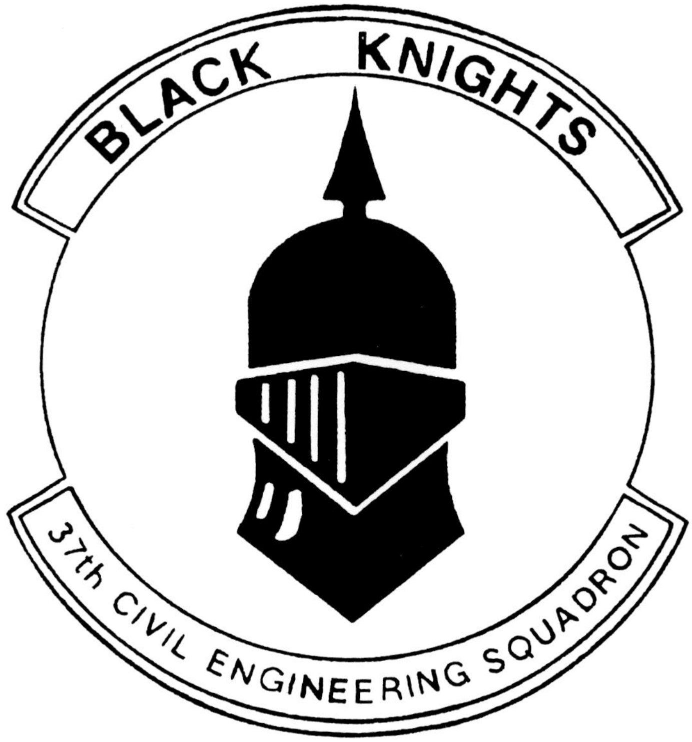

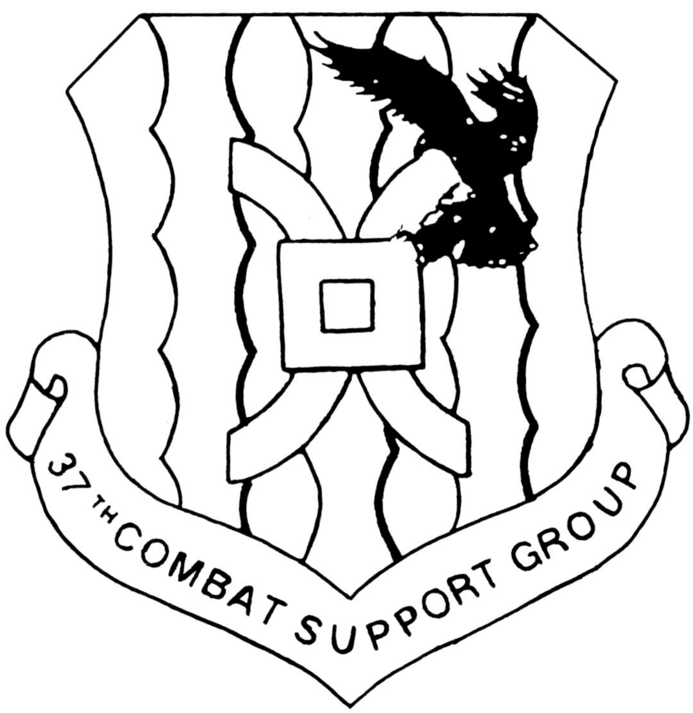

37th COMBAT SUPPORT GROUP

"The design is composed of the colors of the Air Force, with three Air Force yellow segments and a total of seven segments forming the group designation. The three yellow segments also represent the three squadrons which comprise the 37th Combat Support Group. The millrind, the center on which the millstone turns, is symbolic of the faithful service of the group, which acts as a service center for the maintenance of smooth operation of combat units. And the added black Nighthawk, alighting on the millrind, signifies the support given to it to fulfill the worldwide commitments of the F-117A Stealth Fighter."[10]

37th COMPONENT REPAIR SQUADRON

"The Nighthawk symbolizes strength and commitment to the mission of the Wing. The Nighthawk is threaded in golden brown outlined in black, with eyes of red. The disk is divided diagonally from dexter base to sinister chief into two colors. On the top is Air Force yellow to signify the setting sun and enduring excellence. On the bottom is Air Force blue to signify our sky and the Air Force's theater of operations. On the bottom of the Air Force blue, located below the Nighthawk, is a black gear which symbolizes a mechanism that denotes our intermediate industrial capability and strength. The black color denotes the security classification of our program's technology. The atom design 3/4 in the field of Air Force yellow symbolizes technology and our vast potential, which is made up of many diversified shops. The red color signifies the power and capability of this technology. The scroll will be gray with black lettering. The border of the disk and the scroll are black."[11]

Regular and subdued versions of this patch were produced.

Special thanks to Chief Rick Phillips for donating this patch.

37TH EQUIPMENT MAINTENANCE SQUADRON

"A black Nighthawk, shown in flight, displays the swift and silent nature of the F-117A Stealth Fighter as it approaches its intended target. White highlights around the hawk suggest the aircraft's spectral nature. The ultramarine blue background represents the sky, the primary theater of the Air Force's operations. A waxing moon and stars denotes the Air Force's ability to operate in day and night skies. The hands symbolize the care that equipment maintenance technicians take in their work with the aircraft. Five stars refer to the individual branches that contribute to the squadron's equipment maintenance capability. The hawk's red eye denotes its ability to find and track its prey."[12]

Regular and subdued versions of this patch were produced.

Special thanks to Chief Rick Phillips for donating this patch.

37th MEDICAL SQUADRON
"Red (Heraldic Gules) used in the caduceus, the symbol of the medical profession, represents the blood of life. The yellow symbolizes the Sun and its illumination. It is used to depict the rising of the sun, signifying the daylight portion of a 24 hour mission/ capability. Additionally, the stars on the night portion of the emblem are yellow to show their illumination at night. The color brown is for the Earth. Brown is utilized to show the mountainous topography of the region of the Tonopah Test Range in which the 37th Medical Squadron is located. The light blue is used to depict the sky and is used to show the daylight portion of the unit's 24 hour mission/ capability. Black is used to portray night and is used to show the night time portion of the unit's 24 hour mission/capability. And silver, representing purity, is used to color the staff portion of the caduceus."[13]

37th SECURITY POLICE SQUADRON
"The emblem is symbolic of the squadron's mission and history. The ultramarine blue background alludes to the sky, the primary theater of the Air Force's operations. The three red bars are symbolic of the unit's service in South Vietnam. The three black darts represent F-117A Stealth Fighters of the three tactical fighter squadrons of the 37th Tactical Fighter Wing the squadron is tasked to defend. The black chess knight, the most maneuverable of pieces in the historic game, suggests the maneuverability of the unit to deploy anywhere with the aircraft and the quick response of security forces to any threat against wing resources during daylight or darkness."[14]

IRAN TEAM - PLATY SHOP F-117A

The Inspect and Repair As Necessary (IRAN) Team was put together by Lt. Colonel Jay Gaskin, 37th Deputy Commander of Maintenance, in 1990. The IRAN Team's charter was to indoctrinate the squadron's maintenance personnel in the proper inspection and repair procedures for the F-117A. More specifically, the team was responsible for the upkeep of the aircraft's exhaust system, known as the Platypus.

Mike Manuel, a LADC Field Service Rep on location in Tonopah, was involved in the training of the team members and provided input for the design of this patch intended to represent this team.

The shape of the patch represents that of the F-117A, while the stars and moon signify the aircraft's nighttime operations. Hidden at the bottom of the patch is a black silhouette of the top view of the F-117A. Encrypted in this area is the word "Platy," revealing the patch's affiliation, "Platy Shop F-117A." The platypus depicted on the patch is clutching a mason's trowel, symbolic of the people who worked on the Platys, and their tools.

Special thanks to Mike Manuel for donating this patch.

Other Related Support Group Patches

DON'T ASK! ? NOYFB

This patch dates back to the production days of the F-117A Program. As a F-117A was rolled off of the production line, it was readied for delivery to the Air Force via a C-5 Galaxy. Soon after the first mission, this patch was designed. The C-5 crew would strip all insignia and patches from their flight suits prior to departing on their mission and Velcro this patch onto their flight suits in place of their squadron patch. This was done in an effort to maintain their anonymity.

The question mark is symbolic of the cargo, the Top Secret F-117A, of whose existence only those involved in the project knew. "Don't Ask! None Of Your F#&%ing Business" exemplifies the crew's convictions at preserving the secrecy of the Stealth Fighter Project.

The late-night excursions performed by the crew in delivering the F-117As are depicted by the crescent moon with the black background. The shape of this patch is derived from that of their "C-5A Galaxy / 22 MAS" squadron patch removed during these delivery flights.

It is unknown as to how many of these patches were produced by the squadron. However, the author had 200 copies of the Don't Ask patch produced on 21 April 1992 due to the demand within Flight Test for this patch. These 200 reproductions can be identified by the improved quality of lettering due to the use of computer-controlled stitching machines.

Special thanks to Ron Mosher for donating the C-5 Galaxy/22 MAS patch for my collection.

SPECIAL PROJECTS - SEMPER EN OBSCURUS

This patch was used as a symbol for the Special Projects Office (SPO) responsible for overseeing the maintenance and support for the F-117A Program. The SPO operates under the United States Air Force's Sacramento Air Logistic Command (SM-ALC/QL).

The Latin phrase "Semper En Obscurus" translates to "Always in the Dark." This phrase in conjunction with the mushroom is symbolic of the secrecy under which this organization operated.

Special thanks to Laura King for this patch.

PS 77 MOD F-117A

The PS-77 Mod Center patch was designed by Master Sergeants Scott Rutledge and Bill Manning. Scott and Bill were assigned to the F-117A Depot Mod Center in Palmdale, California, as USAF Quality Assurance personnel.

The Mod Center, also known as just PS-77, is responsible for performing configuration update modifications to the F-117A. These mods could pertain to a variety of items ranging from avionics to airframe components.

The view of the F-117A depicted was chosen because it is the most recognizable view, thus dictating the shape of the patch. The black color is representative of the F-117A. The blue and gray colors were selected because they provided a nice contrast while not attracting attention. This color scheme philosophy is symbolic for the F-117A's stealth characteristics.

Originally, 500 patches were produced in January of 1991. Another 300 were made in August of 1993 just prior to the Skunk Works' 50th Anniversary picnic.

Lockheed Product Excellence Patches

HISTORY OF PRODUCT EXCELLENCE

Lockheed's support of a formal concept which promotes quality products and workmanship dates back to the 1960s. During the early days of Kelly Johnson, the famous founder of the Lockheed Skunk Works, this concept was known as Quality Circles. More recently, similar concepts such as the Deming Award and the Malcolm Baldridge Award have been given to companies in recognition of their quality workmanship/performance and ability to provide a defect-free product. All of these ideas contributed to the formation of Lockheed's Product Excellence program.

The philosophy of Product Excellence was developed by Lockheed-California Company (CALAC) President Dick Heppe and Bob Murphy, Director of Operations in 1982 on the foundation of Quality Circles. Bob assigned Mike D'Amico the responsibility of implementing this philosophy throughout the Operations and Quality Assurance branches in Buildings 309 and 310. Advanced Development Projects (ADP) formally incorporated this philosophy into what became known as Product Excellence Program in 1983 and 1984. The policies set forth by this philosophy were written for the OPS MOD (Manufacturing Operations Directives) and QA's (JI) Manuals.

"The first concern of Product Excellence was to instill the ADP philosophy of Kelly Johnson's "Quick, Quiet, and Quality," along with improving schedule and budget. An integral part of this philosophy was teamwork and pride, and an employee recognition program."

"In the progression of developing teamwork and pride, various Product Excellence patches were designed. Each geographic location within the Skunk Works began designing and presenting patches to their employees. By this time, the Product Excellence Program had expanded company-wide, and Mike D'Amico was assigned to CALAC's President, Dick Heppe."

"Richard Bronson was promoted from Buildings 309, 310, and 351, and given the responsibility of implementing Product Excellence in ADP for Bob Murphy."

Greg Cooper and Bob Driver of Lockheed's illustration department contributed to the artistic development of most of the Product Excellence patches. Additionally, the following individuals were members of the Operations' Product Excellence organization and were instrumental in collecting the ideas of the different organizations on which these patches were made: Mary Ann Vernola (Bldg 199), Fred Geiger (Plant C-1), Sharron Hudis (Bldg 371), Deborah Taylor (Bldg 82), Bob Paci (Bldg 304 and 357), Ken Haworth (CME), Lou Armbrecht (Site 2), Jerry Knotts (Site 7), and Dick Anderson (Site 7). Fran Lynch took over for Fred Geiger after he transferred to Lockheed in Ontario, California.

Although the patch campaign was in existence from about 1984 to 1989, many of the designs have origins dating as far back as the 1940s. Information regarding some of these patches may be lost forever. Where possible, as much information has been included. Although many of these organizations supported numerous Skunk Works' programs, in essence they supported the F-117A, whether it was through lessons learned or through their ability to get the job done. To these individuals, I dedicate this section.

Special thanks to Richard Bronson for contributing to this passage.

BURBANK PLANT B-6 LAYOUT

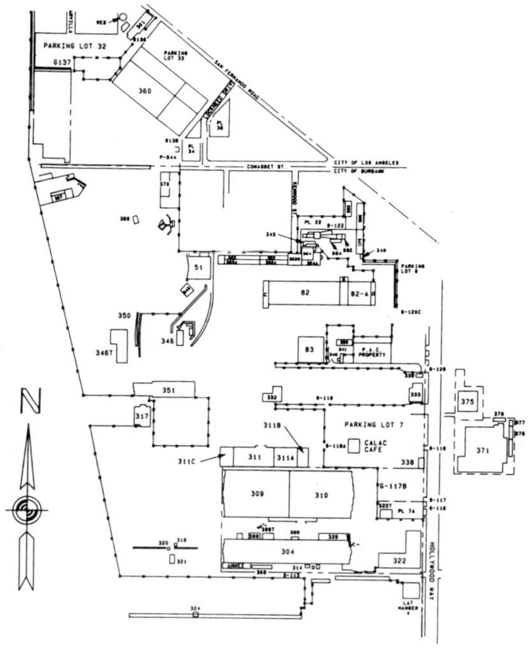

ADP PRIDE - PRODUCT EXCELLENCE
"This patch was first issued under Bob Murphy's direction, but was used extensively throughout Site 2 and 7 (in the USAF's Palmdale Plant 42 complex) before being used on the F-117A Program. The patch was widely adopted throughout ADP because it was the first ADP branch patch, not specific to any one program. The patch was given to employees, customers, and vendors in appreciation of their contributions to ADP." The patch was produced in 1984/85.

This generic, white-washed patch incorporated Kelly Johnson's philosophy: "Quick, Quiet, Quality." Prior to its use in a patch, this design was used by Joe Rinella as an unofficial logo on all customer supply procedures from 1979 to 1980.

A second version of this patch was produced with a silver/gray cloth background instead of the silver embroidered background.

Special thanks to Linda Pearlman and to Richard Bronson for trading me these patches, and to Richard Bronson for contributing to this passage.

CALAC ADP TEAM 1

Prior to the disclosure of the F-117A, it was understood by those among the ranks that "Team 1" was in reference to Burden Center 20, the F-117A production line. This rectangular patch was the first of several patches designed specifically for members of the F-117A production line to help develop teamwork and pride in this rapidly expanding workforce. CALAC was given the go ahead for the F-117A Program. ADP, embedded in CALAC, was known for its revolutionary designs and products, and was given the task of making the F-117A a reality.

Greg Cooper and Bob Driver of Lockheed's illustration department were instrumental in the development of several candidate patches. A red, white, and blue theme was finally chosen to symbolize the patriotism of those working on the aircraft. John Stanley, who was the production manager, recalls that his son (John) also worked on these designs and made the final selection of this patch and the Team 1 Award patch.

Special thanks to Bob Paci for trading me this patch.

TEAM 1 AWARD - PRODUCT EXCELLENCE

This triangular shaped patch adopted the color scheme from its predecessor and was also developed by Greg Cooper and Bob Driver. Under the direction of John Stanley, this second-generation patch had to be generic and sanitized of any company affiliation.

The intent of this patch was to provide a means to recognize individuals that supported the F-117A Program, including CALAC employees outside of ADP, customers, and vendors. "Hence, the patch was designed as a "Product Excellence Award" and given out as recognition."

Special thanks to Bob Paci for trading me this patch and to Richard Bronson for contributing to this passage.

TEAM ONE FURTIM VIGILANS

The Team One Furtim Vigilans patch was the predecessor to the Team One Quality Leader patch. The literal translation of Furtim Vigilans from Latin is "Stealthily keeping vigilance." This Latin phrase also has the same interpretation in Spanish. Because this phrase had the same meaning in contemporary Spanish and because the program and its technology was not acknowledged, the customer requested that Lockheed have these patches recalled and destroyed. Of the 1,500 patches that were produced and distributed to the F-117A production personnel, all but a small number were returned for destruction.

A new patch, in which the Latin was removed, was produced to be reissued. This patch incorporated the "Team One Quality Leader" motto and is described below.

TEAM ONE QUALITY LEADER

Team One was the name given to the F-117A production line led by John Stanley. This patch is the third-generation, second version of the Team One patches, with the first version being the Team One Furtim Vigilans patch.

Given to employees working in Buildings 309, 310, 311, and 351, this patch represents a clever compilation of various F-117A patches. Influencing the shield shape of this patch was the Tactical Air Command logo, while the red and black colors of the Baja Scorpion patch provided the dominant color scheme.

One will notice that the griffin has the tail of a scorpion, also borrowed from the Baja Scorpion patch. The head, wings, and talons of a Nighthawk were incorporated into the body of the griffin. The lightning bolt was taken directly from the Blue Streak patch, and the griffin's legs, torso, and arms were incorporated from the body of the coyote in the same patch.

Designed by Richard Bronson during the Product Excellence patch drive, approximately 1,500 were produced. Copies of this patch are available through commercial vendors, however, there are minor variations from the original seen here.

Special thanks to Bob Paci for trading me this patch.

POWER PLANT LAB

This patch represents the group responsible for testing the F-117A's engines and inlets. Its origin dates back to a cartoon drawn by Bill Eddy in 1948. The cartoon depicts a building housing Jet Motors, Inc. The building is flying through the air after breaking loose from its foundation during an engine test. The crumbling backside of the building exposes a jet engine with a technician riding on top of an engine. Hanging onto the outside of the building is another technician. A caption at the bottom reads: "Give her another notch, Doc - She's starting to percolate." This cartoon inspired the final design for the above patch.

As with other building patches produced during the Product Excellence patch drive of 1986, the Building 88 Power Plant personnel were encouraged to submit ideas for the design of their building patch. Henry Gill, Jr., a lead engineer in the Burbank Power Plant Lab, suggested the idea of an engine flying through the air. The technician riding on top of the engine was replaced with the ADP Skunk. Because the Skunk was riding the engine as though the engine were a horse, it was decided to give the Skunk a cowboy look with the addition of a cowboy hat and saddle. One will note that this scenario is similar to the famous scene in which Slim Pickens is riding a bomb in the movie Dr. Strangelove.

It is estimated that 50 patches of this design were produced.

Special thanks to Karrie Young for donating this patch.

304 TEAM - GROUND SUPPORT SPECIAL PROJECTS

A product of the Product Excellence patch competition of Building 304, this emblem was designed by Jim Leatham. One-half of this building was dedicated to Ground Support. Ground Support was the group responsible for constructing the stands that surrounded the F-117A production line tooling. This group also provided tow bars, tugs, run-up carts, and other Aircraft Ground Equipment (A.G.E.) for the SR-71, U-2/TR-1, F-117A, and various other Lockheed programs collocated at Plant B-6 in Burbank.

The other half of Building 304 was occupied by Special Projects. The name for this group came from a team selected from Building 304 personnel to work on SR-71 tailpipes. In the mid 1980s, this team combined with another and became responsible for building rotators, pylons, and other hardware for the Helendale and Rye Canyon anechoic ranges.

Two of Lockheed's trademark logos are depicted in this patch, along with the familiar lightning bolts seen in other F-117A patches. Approximately 200 patches were produced from 1985 to 1986.

Special thanks to Bob Paci for trading me this patch.

LOCKHEED Q-TEAM

Any employee exhibiting quality workmanship above and beyond the call of duty, or one who may have contributed/implemented ideas for cost savings, improved safety, etc., was eligible to receive a Product Excellence award. This design was developed originally as a lapel pin to be given out by the Product Excellence committee to recipients of this Award. Representing a composite of ideas by John Berman and Dave Walker, this design became a central theme for Product Excellence and was developed into patch, decal, and bumper sticker form. The wide use of this logo, which was refined by John Berman, has made Q-Team synonymous with Product Excellence.

Approximately 2,500 of these patches were produced in 1985. Special thanks to Bob Paci for trading me this patch.

BURBANK PLANT C-1 LAYOUT

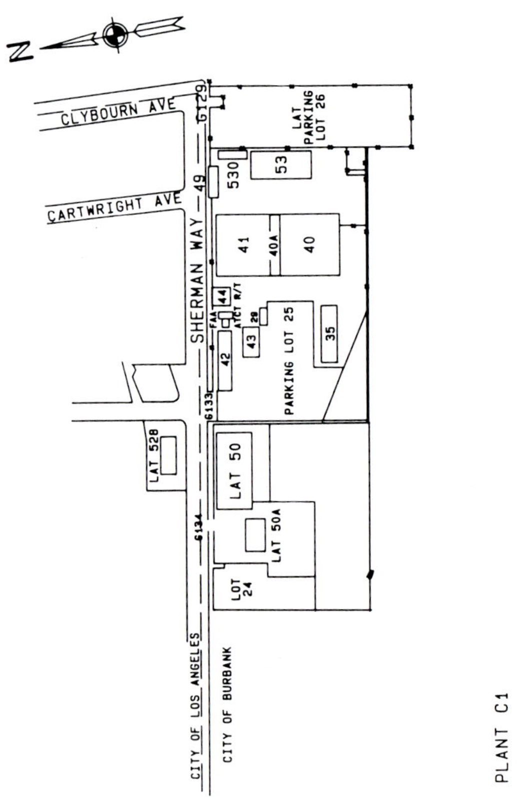

C-1 TEAMWORK

This patch is the symbol for the original ADP machine shop located in Burbank's Plant C-1. C-1 was part of the Skunk Works' complex and had been involved in many of Lockheed's successful programs. This machine shop, dating back to the 1950s, was responsible for fabricating components for the SR-71, U-2/TR-1, and F-117A.

The words Pride, Quality, Creativity, and Teamwork encircling the patch's border were selected to encourage those qualities in the workplace. The arrowheads, indicating that one quality influences the next, made for a nice division between the words. The phrase "Investment in Excellence" is symbolic of the Product Excellence's involvement in the design of this patch.

Designed by Wayne Shaw and Fred Geiger during the Product Excellence patch drive of 1987, about 500 patches and 500 pins were produced.

Special thanks to Wayne Shaw for donating this patch.

QA DEDICATED TO PERFECTION

It was during 1984 that this design evolved into a patch to symbolize the formalization of the Quality Assurance organization from what was once known as ADP Inspection. This was the time frame when the Lockheed Skunk Works adopted the MIL-Q-9858A Quality Program Requirements. The director of QA, Steve Myers, directed Bill Wysong to coordinate the ideas of his coworkers and to come up with a final design. The general consensus was that everyone wanted a design that looked good and whose shape was unique.

The colorful sunburst was Bill's contribution to the patch and was inspired by a design depicted on an early Apollo mission patch which Bill admired. The intertwined Q and A incorporated into this patch had been used as an office logo since 1983/1984. The Skunk represented Lockheed Advanced Development Projects. The ideas from this community design effort were professionally assembled into the design depicted here by Lockheed's Graphic Arts department. This patch has been used as the official logo for the Skunk Works' QA organization since the patch's inception.

Immediately after this patch's design, its image was produced on ballcaps, T-shirts, baseball jerseys, and tie tacks. It is estimated that over 1,200 of these patches have been produced.

Special thanks to Jerry Conner for donating this patch.

FORT

This patch was developed during the Product Excellence patch drive in 1986. The Product Excellence committee contacted George Rivera, department manager of the organization inhabiting Building 357 in Lockheed's Plant B-6 in Burbank. George was asked to have his organization participate in the Product Excellence program by coming up with a patch design to represent the services provided by their group.

George asked his crew to come up with individual designs for their group's patch and offered ten dollars out of his own pocket as the prize for the best design. The design that won, depicted here, was created by Mike Gutierrez. Building 357 was utilized as a bunker for testing armament during the 1940s. The building's outward appearance led to its nickname, "the Fort," a name which has stuck since the 1950s. This nickname was symbolized by an American frontier fort defended by a Skunk dressed in a U.S. Calvary uniform. The Skunk represents Lockheed's Advanced Development Projects as the symbol for the Skunk Works. The outer border of the patch was made to look like a test gauge dial. The dial represents the Functional Tests performed on components from the SR-71 and the F-117A. Components for the F-117A's bomb bay doors and fuel cells were tested in mock-ups located in Building 357.

About 100 patches were produced.

Special thanks to Bob Paci for trading me this patch.

QUALITY SEALANT PROGRAM

The Quality Sealant Program patch was developed for Buildings 309 and 310 employees by Keith Hunter in 1986. During 1985/1986, a major effort was undertaken to eliminate fuel leaks on the F-117A. All fuel leaks were reduced significantly or eliminated during this period. This patch was given to employees who completed sealant training classes as fuel tank sealers on the F-117A.

Approximately 200 of these patches were produced.

Special thanks to Keith Hunter for contributing this patch and passage.

QUALITY FASTENER PROGRAM

The Quality Fastener Program patch was developed by Keith Hunter for Buildings 309 and 310 in 1985. This program patch was developed along with a training course for all structures employees. The purpose of this program was to insure that all exterior fasteners on the F-117A were flush with the skin so that the coatings would adhere without excessive bondo application.

Over 1,000 of these patches were produced.

Special thanks to Keith Hunter for contributing this patch and passage.

ACHIEVEMENTS DEVELOP OUR PRIDE

The Achievements Develop Our Pride patch was developed in Buildings 309 and 310 by Gus Villanueva and Keith Hunter in 1984. Initially developed for the employees working on the F-117A's aft body titanium area, this patch was later expanded to all eight structures departments. This patch was issued to each department completing its work on time, under budget, and with the highest of quality.

Approximately 1,000 of these patches were produced.

Special thanks to Keith Hunter for contributing this patch.

20-02 SOUTHSIDE / TAKING CARE OF BUSINESS

The Structure Divisions were noted for fast, corrective, and preventative action. This design signifies "Taking Care of Business" in a flash and was inspired by Elvis Presley's personal logo. The design was developed in 1986 into a patch and a decal form by Keith Hunter, General Superintendent of Structures, and issued to all structures employees working in Buildings 309 and 310.

Special thanks to Keith Hunter for contributing this decal and passage.

BLDG. 371 OVERHAUL AND REPAIR

Building 371 in Burbank was responsible for performing overhaul and repair work for Lockheed's U-2 and SR-71 Programs. With the advent of the F-117A Program, this group performed overhaul and repair work on the Stealth Fighter as well.

In 1988, a building-wide contest was established to design a logo for Building 371 and its employees. It was decided by building management that the employee submitting the best design would have the privilege of parking inside the gate in the Employee of the Month parking spot. The contest drew interest, and approximately 25 employees submitted designs for review by the Product Excellence committee. Patches and decals with the winning design were produced.

Between 200 and 250 patches were produced with this design depicted here.

Special thanks to Richard Bronson and Tony Aragus for donating this patch.

BUILDING 375 - AERIAL RECONNAISSANCE PROGRAMS

This patch was created during the Product Excellence patch drive to represent the Lockheed Program Office of the Aerial Reconnaissance Fleet of the USAF comprised of the SR-71, the U-2, and the TR-1. The program office was located in Burbank's Building 375 and was responsible for providing support to these aircraft.

A contest was sponsored by Product Excellence in which employees of Building 375 were encouraged to submit entries. The winning entry was developed into the design depicted here by Bob Driver's illustration department. On 26 OCT 87, 1,000 patches, along with various quantities of coffee cups and tie tacks, were ordered with this design.

Special thanks to Bill Crosthwait for trading me this patch.

CME - TECHNOLOGY MANUFACTURING

"During the 1985 time frame, the department/model shop responsible for manufacturing RCS Range Hardware Systems and RCS Test Models was buried in the Chief Manufacturing Engineer (CME) organization. The Model Shop department was under the management of Lee Flanderka, who reported to Robby Robinson. Robby, in turn, reported to Bill Gilchrist, who was CME."

"In 1986, plans were put into effect to design and develop an organizational logo. Requests were made to all personnel for ideas. The selected design embossed the title of the CME with the three major segments of the organization: Planning, Tooling, and Manufacturing Engineering. Also included with the three segments was the ADP Skunk. These elements were laid out on a patch approximately 2 inches by 5 inches, with blue and black lettering and a red border on a white background."

"Approximately 500 patches were produced along with Model Shop coffee cups. Patches were distributed to all CME personnel, and cups were distributed to Model Shop personnel."

Special thanks to Robby Robinson for contributing this patch and passage.

B-82 OFFICIAL MEMBER PRODUCT EXCELLENCE
BLDG 82 ACHIEVEMENT DEDICATION PRIDE

The B-82 Official Member Product Excellence patch was given to all employees assigned to Building 82 while the Bldg 82 Achievement Dedication Pride patch was given to recipients of Awards of Achievement. Both patches were designed by committee. The committee included Ed Banyard, Fran Lynch, Larry Kibble, Art Ziebell, and other personnel from Building 82 in Burbank.

The red, white, and blue color scheme, and the stars separating the words in the outer ring provide a patriotic theme for the Bldg 82 patch. The words "Achievement," "Dedication," and "Pride" were borrowed from an article published in a Lockheed Corporate newsletter.

Sump (fuel) tanks and tailpipes for the U-2/TR-1 were produced utilizing the hydro press, seam and spot welders, and other large machinery located in Building 82. The original idea was to profile a TR-1 on this patch, however, the SR-71 fit the shape of the patch better and was therefore chosen.

Although the SR-71 and the TR-1 were the main programs supported by Building 82, the machine shop and sheet metal shop fabricated F-117A internal frame and external skin components not requiring personnel to have access to the program.

Special thanks to my wife, Marian, for donating these two patches.

FLIGHT TEST - MAKING DREAMS A REALITY

The design of this Flight Test patch developed as a result of a Lockheed Product Excellence contest in 1988. At a time when many patches were coming out of the fielded F-117A Flight Test group, this contest was intended to promote pride and unity throughout all of Lockheed's ADP Flight Test organization, regardless of program affiliation or assignment.

Members of Flight Test were allowed to submit as many design entries as they wanted. The winning design was one submitted by Eddie Velasquez. The idea was to depict Flight Test from the beginning of aerospace to the future. The Skunk, the symbolic trademark of the Skunk Works, makes another rare appearance on a patch. Coffee cups with this design were also produced at the same time that 500 of these patches were produced.

Related Lockheed Patches

COMPOSITE DEVELOPMENT CENTER
The Composites Development Center patch was designed in part by Janet Ancil for the group located in Building 199. This group later relocated to its own building, known as the CDC. The words forming the border of this patch were motivated by the Product Excellence philosophy. The Flying Star was symbolic of Lockheed and is a registered trademark. The number of patches produced of this design is not known.

Special thanks to Edward Warren for donating two of these patches to my collection.

LOCKHEED PALMDALE - 10 FACTORY OF THE FUTURE
The artwork for this design was refined by Greg Cooper's illustration department at the request of Dan Patterson in 1985. Dan was Plant Manager in Palmdale, and aspired to unify the personnel and pull together a solid team.

Efforts were made to motivate employees by giving them something to rally around. Lockheed Palmdale was working on the P-3 and C-5 Programs and was gearing up to work on proposals for the F-22 Program. In one effort to motivate employees, a contest was held to design the first ever Lockheed Plant 10 patch.

The theme of the contest was "Today and the Future." Five employees emerged as winners. The best qualities from each design submitted by these five individuals were then compiled into one patch by Dan and Greg.

The Lockheed Plant 10 Star symbol seen on this patch was developed at this time, as well. The "10" not only represents Plant 10, but is also symbolic of the best on a scale of 1 to 10. Today, the Antelope Valley is the aerospace mecca of the world, being home to Edwards AFB, Lockheed, and other aerospace companies.

Special thanks to Mike Rimiola for donating this patch.

BUILDING 199 - PROFESSIONAL PRIDE
This patch was designed by Tom Doll to represent the Paint Shop crew in cost center 29. Although this crew from Building 199 was known for its work on the SR-71, they also supported the F-117A Program.

Special thanks to Carol Washington for contributing this patch.

IRON WORKS
Not much is known about this patch except that it represented the group in Building 199 responsible for making materials used on the F-117A Program. This patch is believed to have been designed by Tom Doll between 1984 and 1986.

PAINT SHOP - THE FEW...THE PROUD

This patch is thought to have been designed by Tom Doll between 1984 and 1986. The design depicted here was symbolic of the Paint Shop crew consisting of those employees working in cost centers 29 and 42. These two cost centers performed material applications and painting for the F-117A, SR-71, and other Skunk Works' aircraft.

The phrase "The Few...The Proud" and the combat boots were suggested by former Marines (Randy Robb and others) working in the group. The spray gun is representative of the Paint Shop, of which Pete Arroyo was the supervisor.

Special thanks to Pete Arroyo for trading me two of these patches.

FACTORY TRANSPORTATION

The original concept of this patch was initiated by Bret Beck and Dave House of Lockheed's Factory Transportation in the Fall of 1985. Once the design was refined by Bob Driver of Lockheed's illustration department, Bret and Dave assigned colors to it. Two color schemes were developed, one for patches and one for decals, which were used on hard hats. 50 of each were made.

There were three Factory Transportation organizations within Lockheed—White-world, black-world (both in Burbank), and Palmdale. This patch represents the black-world, or ADP, Factory Transportation personnel in Burbank's Plant B-6. The Skunk is symbolic of ADP, and the Flying Star hanging on the crane hook represents Lockheed. Both symbols are registered trademarks of Lockheed.

Many of the crane riggers in this organization were employed by Lockheed since the 1940s and 1950s and were stereotyped as wearing overalls. The overalls worn by the Skunk are in tribute to these dedicated employees. The employees in Factory Transportation fell into the following job classifications:

Power Truckers who operated forklifts and tugs inside the plant and were responsible for moving equipment located on Mercuries.

Aircraft Power Truckers who towed aircraft behind tugs.

Factory Transportation Dispatcher who received the calls for the organization.

Crane Riggers who operated the overhead cranes located inside Lockheed's buildings.

Truck Crane Operators who operated cranes outside of the buildings.

Special thanks to Dave House and Ron Mosher for trading me this patch and decal.

TECHNOLOGY DESIGN

"In 1985 the department responsible for RCS Range Hardware Systems and RCS Test Model Design under the management of Gary Wendt was given the title of Technology Design department. In the 1986-1987 time frame, under the leadership of Bob Retsch, an effort to develop a department logo was established with a mini-contest for all personnel to come up with a design."

The winning design depicting a stealth-type RCS model being tested on a range at night was submitted by Page Jennings. Page incorporated an idea proposed by Brad Althouse to include a series of gear teeth across the bottom, mating with teeth on the "C" in LASC. These gears are symbolic of the drive gear and ring gear assembly of the model positioner.

Coffee cups were printed with this design, along with the 220 patches produced.

Special thanks to Gary Wendt for donating this patch and contributing to this passage.

SITE 2 - ADP'S FINEST

In November of 1988, a contest was held to design a logo to represent the operations at Site 2 of the Air Force's Plant 42 in Palmdale. As an incentive to get employees to respond to this contest, a jacket was offered as the prize for the winning entry. Approximately 20 entries were submitted, and the design chosen was the one submitted by John Homsy, a Flight Test Instrumentation Technician.

The patch depicts the two programs with operations at Site 2, the U-2/TR-1 and SR-71. The Skunk, the registered trademark of LADC, is symbolic of the manufacturer of these airplanes, and the globe represents the vast area in which these reconnaissance aircraft operate.

A 3.5" and a 4.5" version of this patch was produced.

Special thanks to Al Heinrich and John Homsy for trading me this patch.

BUILDING 636

Some of Building 199's personnel in Burbank began relocating to Palmdale's Building 636 in 1989. As a result, a competition to design a Building 636 patch was initiated. This patch is a composite of two designs, one submitted by John Becker and the other by Tom Doll.

John was inspired by a quotation on a picture hanging in 636 and shortened "Some people look at a limit and see an end. Others perceive a new beginning..." into "On the edge of tomorrow." Tom's pictorial depicts a futuristic aircraft originating from Palmdale, leaving the Earth and heading for a star.

It is estimated that 500 to 600 of these patches were produced to represent this group, which was responsible for providing composite components for the F-117A Program.

Special thanks to Ed Beurer for donating this patch.

BUILDING 637

The C-1 machine shop was moved to Building 637 during the relocation from Burbank to Palmdale. The original C-1 patch became outdated since the machine shop was now located in Plant 10. The production manager, Larry Kibble, realized this and designated a committee to redesign their patch to meet the current situation.

The goal in choosing a shape for the emblem was to avoid using the typical circular design evident in other associated patches. The shape which was finally chosen was symbolic for panels fabricated by this facility and thus represents a generic part. The Skunk was carried over from the C-1 patch to indicate continued association with the Skunk Works, while the colors of the patch were borrowed from the color scheme of Building 637. The facility became known as the Metallic Fabrication Center because it provides machine shop and sheet metal fabrication services.

About 675 of these patches were produced in October of 1989, along with various quantities of tie tacks and coffee mugs.

Special thanks to Wayne Shaw for donating this patch.

LOCKHEED RYE CANYON

The raccoon with the green background on this patch depicts the atmosphere at Lockheed's Rye Canyon Facility. This location provided a pleasant working environment for the Lockheed employees, and especially for the gate guards whose job required them to spend most of their time outdoors. The gate guards were visited each night by raccoons living in the area, and soon a friendship was established with these night animals for which Rye Canyon was best known.

This patch began as a collection of ideas among the Rye Canyon guards and was designed to be unique to LADC and, more specifically, Rye Canyon. In keeping with the original Lockheed Security patch, the shape and yellow color were incorporated into the Rye Canyon Patch. The SR-71, TR-1, and F-117A portrayed on this patch represent the unique LADC operations at Rye.

Bob Nixon collected these ideas and drew the initial concept for this design. This patch was worn by the guards from 1990 until 1993. In September of 1990, 510 patches were produced.

RYE CANYON PLANT LAYOUT

PLANT 2

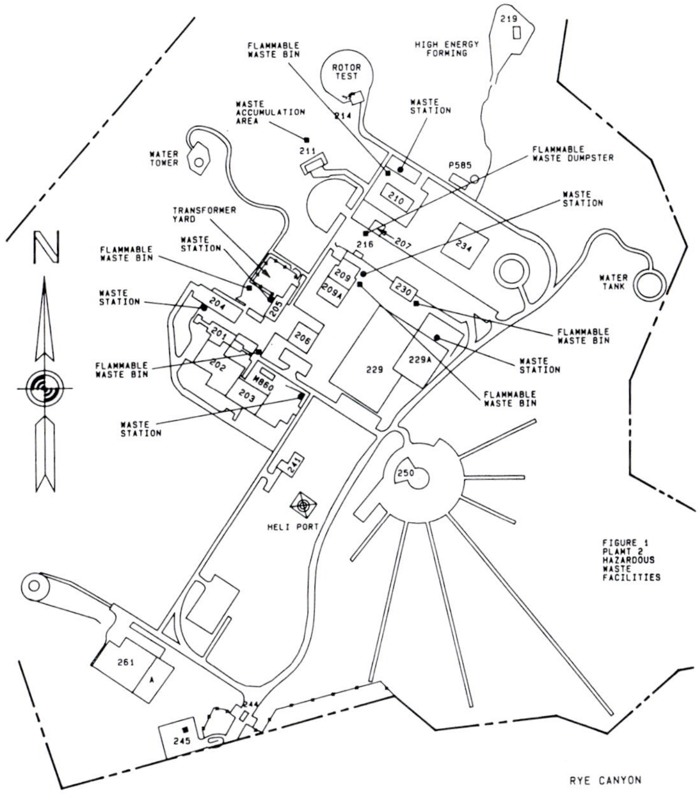

LOCKHEED SKUNK WORKS PALMDALE
LOCKHEED MARTIN SKUNK WORKS PALMDALE

The patch seen here is an updated version of the Lockheed Palmdale - 10 Factory of the Future patch. This re-creation was initiated by Jim Grose of Lockheed's Fire Protection organization in February of 1994 to commemorate LADC's exodus from Burbank to Palmdale.

The intent of this patch was to welcome the former Burbank employees to LADC's new headquarters. The original Plant 10 symbol seen in the earlier version was now replaced with the Skunk Works' famous trademark, and the rest was altered to indicate that the "Factory of the Future" was now here in "Palmdale."

In Jim's preliminary sketches, "Lockheed" across the top was replaced with "LADC." However, it was later decided to leave that portion of the patch unchanged to remain consistent with the earlier version.

An order of 200 was placed by Craig Wilsman of the Lockheed Employees Recreation Club on 10 JUN 94 and are available for purchase through the LERC.

The patch was later updated to reflect the 15 MAR 95 merger between Lockheed Corporation and Martin Marietta.

LOCKHEED SKUNK WORKS
LOCKHEED MARTIN SKUNK WORKS

"The artwork upon which this patch was produced was designed by Mike Whalen of the Lockheed Advanced Development Company Special Arts group in January 1994. Dick Abrams, then Director of LADC Flight Test, admired the logo and decided to have a patch produced in limited quantity for distribution to all Flight Test personnel as a gift from him."

"Requests for quotes and samples were sent to various vendors on March 9, 1994. On May 13, Emblem Enterprises of Van Nuys, California, was selected to provide two samples, and a down payment was placed on a production order of 400 by Dick Abrams."

"The samples, one gray and one black, arrived. The black one was selected for production and the order was placed. The one of a kind gray one remains in possession of Jim Fitzgerald, who handled the paperwork."

"Dick Abrams was taken seriously ill while participating in the Society of Flight Test Engineers' National Symposium in Washington, D.C., on August 2nd, 1994. The production order of 420 patches arrived at Lockheed on August 10 and, ironically, Dick passed away that night. His wish that each member of the LADC Flight Test team receive a patch was carried out the following week by his staff. The remaining patches are being sold within Lockheed."

The patch was later updated to reflect the 15 MAR 95 merger between Lockheed Corporation and Martin Marietta.

Special thanks to the late Dick Abrams and to Jim Fitzgerald for giving me this patch, and to Fitz for contributing this passage.

49th Fighter Wing Patches

49th FIGHTER WING - FIGHTIN FORTYNINERS
Initial transfer of the three F-117A squadrons (the 415th, 416th, and 417th) of the 37th Fighter Wing to the 49th Fighter Wing in Holloman AFB, New Mexico, began with the arrival of four Stealth Fighters from Tonopah Test Range, Nevada, on 09 MAY 92. This relocation was completed on 07 JUL 92. The 49th has a history dating back to 1940, and many of its members throughout the years have held the title of "ACE" and have the distinction of numerous combat kills.

TUTOR ET ULTOR
"The heraldic beauty of the insignia is highlighted by the stars of the Southern Cross, which denote the wing's World War II campaign credits in the South Pacific. The winged knight's helmet reflects the unit's long and distinguished history of military engagements. The bolt of lightning represents the powerful, instantaneous striking power inherent in the 49th Fighter Wing. The Latin inscription TUTOR ET ULTOR translates into the meaningful motto I PROTECT AND AVENGE, and is an ideal reflection of the wing's richly earned nickname, THE FIGHTING 49ers."[15]

7th FIGHTER SQUADRON
The 417th Fighter Squadron of the 37th was the last of the original F-117A squadrons to be redesignated into the 49th Fighter Wing. On 02 DEC 93, the 417th was officially redesignated as the 7th Fighter Squadron. The Screamin' Demons is the unit responsible for providing initial qualification training for pilots in the F-117A. Instructors fly chase in T-38 aircraft and observe the student pilot's performance after numerous hours of training in the flight simulator have occurred. Upon successful completion of flight training, pilots are assigned to one of the operational Stealth Fighter Squadrons.

8th FIGHTER SQUADRON - THE BLACK SHEEP
On 30 JUL 93, the 416th Fighter Squadron of the 37th was redesignated as the 8th Fighter Squadron, keeping with General McPeak's Historical Heritage and Realignment Policy, to preserve the rich heritage of the 8th Fighter Squadron. The Black Sheep is the first of the two operational F-117A Stealth Fighter Squadrons and provides a unique combat capability of providing precision guided weapons delivery against heavily protected targets while remaining undetectable by radar.

9th FIGHTER SQUADRON - THE IRON KNIGHTS / THE FLYING KNIGHTS

On 30 JUL 93, the 415th Fighter Squadron of the 37th was redesignated as the 9th Fighter Squadron, keeping with General McPeak's Historical Heritage and Realignment Policy, to preserve the rich heritage of the 9th Fighter Squadron. The Flying Knights is the second of the two operational F-117A Stealth Fighter Squadrons and participates in exercises around the world, along with participation in airshows and public displays.

NIGHT HAWKS - 49th OPS SUPPORT SQUADRON

The original unit, the 49th Airdrome Squadron, was disbanded on 01 APR 44. On 01 NOV 91, the unit was "reconstituted, and redesignated 49th Operation Support Squadron," and activated on 15 NOV 91. Afterwards, this emblem was developed and approved on 09 JUN 92.

"Blue and Yellow are the Air Force colors. Blue alludes to the sky, the primary theater of Air Force operations. The yellow zia sun symbol refers to the excellence required of the Air Force personnel. The nighthawk grasping several lighting bolts represents the aircraft and reflects the squadron's support of the wing's nighttime delivery of air-to-ground precision munitions mission. The winged helmet suggests the Air Force's role in national defense. The sword and lighting flash stand for the technology and equipment used to ensure that defense and reflect the squadron's efforts in support of defense."[16]

AIR COMBAT COMMAND (Hidden Symbology)

This patch is an unofficial rendition of the ACC patch worn by the members of the 49th Fighter Wing. This patch is larger than the official ACC patch and has hidden symbology embroidered into the background. Can you find all four?

SCREAMIN' DEMONS - STEALTH DRIVER
This patch is believed to be an unofficial patch that has been seen being worn by F-117A pilots of the 7th Fighter Squadron. The name sums it up!

BLACK SHEEP - STEALTH DRIVER
Like the previous patch, this too is believed to be an unofficial patch that has been seen being worn by F-117A pilots of the 8th Fighter Squadron.

FLYING KNIGHT - STEALTH DRIVER
Like the previous two patches, this too is believed to be an unofficial patch that has been seen being worn by F-117A pilots of the 9th Fighter Squadron.

Footnotes

[1]Department of the Air Force, Headquarters United States Air Force Historical Research Center, Maxwell Air Force Base, Alabama.

[2]Office of History Headquarters, 37th Fighter Wing, Twelfth Air Force, Tactical Air Command, *History and Lineage of the F-117A Stealth Fighter Organizations - Special Study HO-91-2* (December 1991), P. 53.

[3] "History", P. 59.

[4]"History", P. 6.

[5]"History", P. 14.

[6]"History", P. 21.

[7]"History", P. 28.

[8]"History", P. 49.

[9]"History", P. 41.

[10]"History", P. 38.

[11]"History", P. 51.

[12]"History", P. 41.

[13]"History", P. 45.

[14]"History", P. 43.

[15]Office of History Headquarters, 49th Fighter Wing, Air Combat Command, *A History of the 49th Fighter Wing*, Holloman Air Force Base, New Mexico.

[16]Office of History Headquarters, 49th Fighter Wing, Air Combat Command, *Lineage and Honors History of the 49th Operations Support Squadron*, Holloman Air Force Base, New Mexico.

Glossary

A/C	Aircraft
ACC	Air Combat Command
ADA	Angle of Attack
ADP	Advanced Development Projects (A.K.A. "The Skunk Works)
A.G.E.	Aircraft Ground Equipment
AOL	Aircraft Operating Limit
APU	Auxiliary Power Unit
Bandit	F-117A Pilots' callsign
BDU	Bomb Dummy Unit
CALAC	Lockheed-California Company was formed as a division of Lockheed Corporation in the 1950s and used the acronym CALAC until it was reorganized in 1987.
CME	Chief Manufacturing Engineer
CTF	Combined Test Force of Lockheed and USAF personnel
DFMT	Depot Field Maintenance Team
EPU	Emergency Power Unit
FOT&E	Follow-on Testing and Evaluation
FSR	Field Service Rep
FTE	Flight Test Engineer
GBU	Guided Bomb Unit
GIB	Guy In Back. Non-Pilot crew member
INS	Inertial Navigation System
IRADS	Infrared Acquisition and Designation System
IRAN	Inspect and Repair As Necessary
LADC	Lockheed Advanced Development Company. When the Skunk Works became a separate company during the corporate restructuring in May 1990, it was given the name LADC.
LASC	Lockheed Aeronautical Systems Company. LASC was the Lockheed company in Burbank in which the Skunk Works was imbedded from 1987 to 1990.
NWS	Nose Wheel Steering
OFP	Operational Flight Program
QA	Quality Assurance
RCS	Radar Cross Section
RTU	Replacement Training Unit
S/N	Serial Number
SAC	Strategic Air Command
SPO	Special Projects Office of the USAF
TAC	Tactical Air Command
Troika	A heavy Russian sled pulled by a team of three horses. Also, an administrative council consisting of three people.
Tweak	Electrician/Avionics Technician
USAF	United States Air Force
WDC	Weapons Delivery Computer
WSC	Weapons System Computer

Additional Acknowledments

Collecting these patches and/or the associated information may not have been possible without help from the following individuals:

Tony Aragus
TSgt Tim Bear (DET 1, 57th WG)
Vic Bender
Gary Coots and daughter, Stephanie
Bill Crosthwait
James Farrow
Nelson Hicks
Greg Hill
Kieth Hilton
Mark King
Yancy Mailes
Terry McMahon
Ron Mosher
Rick Phillips
Marty Roshalt (International Insignia)
Patti Rush
Ron Seres
Bobby Wall
Curt Wallis

Copyright by LMSW 1998.